COMPUTER ORGANIZATION
Basic Processor Structure

COMPUTER ORGANIZATION
Basic Processor Structure

James Gil de Lamadrid

CRC Press
Taylor & Francis Group
Boca Raton London New York

CRC Press is an imprint of the
Taylor & Francis Group, an **informa** business

A CHAPMAN & HALL BOOK

CRC Press
Taylor & Francis Group
6000 Broken Sound Parkway NW, Suite 300
Boca Raton, FL 33487-2742

© 2018 by Taylor & Francis Group, LLC
CRC Press is an imprint of Taylor & Francis Group, an Informa business

No claim to original U.S. Government works

Printed on acid-free paper
Version Date: 20180131

International Standard Book Number-13: 978-0-8153-6246-3 (Hardback)
International Standard Book Number-13: 978-1-4987-9951-5 (Paperback)

Visit the Taylor & Francis Web site at
http://www.taylorandfrancis.com

and the CRC Press Web site at
http://www.crcpress.com

To my father and mother, who infused me with the conviction that there are very few human activities more important than pursuing knowledge.

Contents

Introduction and Remarks

INTRODUCTION

This book tries to answer the question, "How is the processor structured?" This question leads to a second question: "How does the processor function in a general-purpose computer?"

The answers to these questions can be quite complex and quite involved, but the answers to these questions do not need to be all that complex. The complexity of the answers to these questions should be appropriate for the audience to which the responses are addressed. If you are addressing a processor designer, of course the answers must be very detailed. However, if you are addressing a layman, the answers would be fairly simple and abstract.

This book is intended to be used in a computer science curriculum. So, our audience is assumed to be computer science undergraduates, or lower-level graduate students. As such, the answers we supply to our motivating questions do not have to be nearly as detailed as the answers we would give to a potential processor designer, nor should they be as simple as the answers given to a layman.

The pedagogical question that drives the content of this book is, "What is the simplest explanation of a processor you can give to a student of computer science; an explanation that will not overpower the student with information, during the learning process, and yet is sufficiently complete so as to serve the student in their career?" In this book, we believe that we have found the sweet spot between too much, and too little information.

Our choice of topics and depth of coverage in this book are based on a couple of decades of teaching experience. Having taught computer organization, and architecture, over the years we have settled on a set of topics that, we believe, is the essence of the field. The set of topics is small enough so that all of the topics can be taught in a single semester course.

WHY STUDY COMPUTER ORGANIZATION AND ARCHITECTURE?

There are two topics that are often taught as part of the computer science curriculum: computer organization and computer architecture. Collectively, we will refer to these two topics as computer systems. Computer systems is a little out of line with most other computer science topics, which are mostly concerned with software.

Usually, the job of describing computer hardware is split into two levels. Computer organization concerns itself with low-level circuity, or, as we might say, *how* the computer computes. Computer architecture concerns itself with higher-level devices, and how these are manipulated by software, or, as we might say, *what* the computer is capable of computing.

This book concerns itself with both computer organization and computer architecture, with an emphasis on computer organization. We cover some computer architecture with our material on machine language programming.

Often enough, we are asked by students, and sometimes colleagues, what relevance computer system courses have to computer science. This question usually comes from a view that computer science is the study of software only, and that learning how the computer works on a hardware level is irrelevant. But, there are at least two responses to this question that establish the importance of computer systems in the computer science curriculum.

The first response takes issue with the view of computer science as concerned with software only. There are several fields that are embraced by the computer science field that work on the frontier between hardware and software, and in which the worker must have an understanding of hardware function, as well as software skills: robotics, embedded systems, and even operating systems.

The second response takes issue with the perception that if you work only with software, hardware knowledge is irrelevant. Imagine that you are hiring a computer scientist for software work. If faced with two candidates, the question is which would you view as preferable: a candidate with a solid understanding of how to program a computer, as well as an in-depth understanding of how the computer behaves, or a candidate with only software experience? You would probably view the more in-depth knowledge as an asset. Knowing how a computer functions does help, immensely, in software development. Many of what would seem unexplainable software behaviors often become clear with hardware knowledge, giving the worker an advantage in the testing and verification of software. Also, notice that knowing how hardware functions can often also lead to smarter design decisions when building software systems.

USING THE BOOK

This book has been designed as a teaching aid, to be used in a one-semester course on computer systems. It covers most of the essential topics in computer organization, and few topics in computer architecture.

The course with which this book would be used, would be aimed at students at an undergraduate level. The course for which we use this book is taught for graduate students. The course in which we teach this curriculum is used to insure that incoming graduate students all have exposure to a common core, of which this course is one component.

To use the book for a single-semester course, it is possible to cover almost all material in the book. You can start with Chapter 1, which gives the student perspective on the interaction between hardware and software. This chapter takes the reader through the process of getting a program to run. It starts with creating the software, compiling and assembling the software, loading it into memory, and running it. It then briefly explains how executing instructions results in operations in digital circuitry. After this overview, we start detailing the processes described.

Chapter 2 presents the mathematical basics required in the rest of the book. In particular, we present material on Boolean algebra and the binary number system.

In Chapter 3, the basics of digital circuitry are discussed. We are taken through the basics of combinational circuits. Then, we examine sequential circuits. This is followed by Chapters 4 and 5. In Chapter 4 we talk about the bus communication architecture, used in many computer systems. A brief discussion on interfacing with peripheral devices is included. Chapter 5 talks about the RTL level of circuity. In this material we describe the building blocks of the processor.

Chapter 6 is a discussion of machine language, that finishes off the preparation for processor design. We talk about the different processor architectures, in terms of the number of operands in the machine instructions, from 0-operand stack machines to 3-operand register machines.

It is possible to cover Chapters 1 through 4, and most of Chapter 5, by half-way through the semester. This gives the student a good understanding of the preliminary information required to understand processor design.

In the second half of the semester, you could cover Chapters 7 through 9. In Chapter 7 we design a processor. The processor is designed as an algorithmic circuit, starting with the data path, and finishing with the control unit. A relatively simple register-implicit machine is designed; simple enough so that details do not lead to confusion, yet with enough complexity so that the reader will see it as useful.

In Chapter 8, we talk about ALSU design and computer arithmetic. The usual operations of addition, subtraction, multiplication, and division are covered, for both integer types and floating-point.

Chapter 9 discusses micro-controlled processors. We redesign the control unit for the same processor covered in Chapter 7.

Chapter 9 concludes what we would think of as fundamental computer systems information. In Chapter 10, we briefly consider several more advanced topics. When we teach this material, this information is presented at the very end of the semester, as time allows. Usually we end up giving a short overview of these topics only.

OUTSIDE RESOURCES

There are a couple of resources available to the student and instructor for enhancing the material presented here.

- The solutions manual. This is available to the instructor, through the publisher. It includes answers to all exercises in the book.

- The BRIM Simulator. The machine language BRIM (Basic Register Implicit Machine) is used throughout the book, as the interface to the main architecture presented. A simulator for the machine level of this machine is available. It includes an assembler and an emulator. An executable build exists for the Linux and Windows platforms. It can be run on the Macintosh, by creating a build from the source Python scripts.

There are other tools that can be used with the book, when exploring some of the topics. In particular, when working with digital circuitry, it is often useful to use a circuit simulator. With the ready availability of circuit simulators, we invite you to choose your favorite one, and use it in your instruction. Our favorite is Logisim.

Our approach is to do digital circuitry in simulation, but you could opt for something more substantial, like a hardware lab. Or, as we do in our undergraduate course, you could run an FPGA lab. We include a small section on Verilog programming, which can enable the reader to program for a variety of RTL simulators, such as Modelsim, or program for synthesizers.

If instructors wish to introduce a group project into the course, they could ask the students to build the BRIM machine as a class project. With a whole class working on the processor, this project could be completed in Verilog or VHDL, in a semester's time.

Overview

CONTENTS

Often students begin their training in the field of computation by learning to program a computer. In this experience the student learns how to construct a program in some *high-level language*. The student learns how the syntax of the language, and the semantics, work together to form a description of a computation. The student also learns that this description can be run on a computer to perform the computation. Many of us then start wondering what kind of magical machine is capable of performing our computation, given only a simple description.

The truth of the matter is that computers are indeed magical, in the sense that they can do amazing things. However, they are less than magical, although still amazing, when we examine how they work. Two areas that examine the operation of the computer are called *computer organization* and *computer architecture*. The distinction between computer organization and computer architecture is probably best described in terms of two other concepts: *implementation* and *interface*. Computer organization is the study of the implementation of a computer. That is to say, the hardware and circuitry out of which the computing system is built. Computer architecture studies the structure presented to a program, that can be used to perform a computation.

Computers are made up of many separate devices. To understand how a computer performs a computation, probably the most interesting device to examine is the *central processing unit* (CPU), which is often simply called the *processor*. This is the device that actually executes a program, and is the focus of this book. In this section we directly examine the question of how a program performs a computation, by giving an overview of the structure of the computer system. In this discussion the structure is presented as a set of layers, one built on top of another. We begin our discussion at the top level, the high-level language program which has been written, and work our way down to the bottom level, which consists of digital circuitry.

1.1 HIGH-LEVEL, ASSEMBLY, AND MACHINE LANGUAGES

So, a user writes a program in a high-level language, with the intent of running the program on a computer. Let us analyze that statement. We begin by discussing what is meant by a *high-level language*.

1.1.1 High-Level Languages

Programming languages are used to describe a computation that is to be executed on a computer. Languages are classified by level. The idea is that low-level languages are closer to the hardware of the computer, and high-level languages are further from the hardware, and more abstract. That is to say, a low-level description of a computation would describe the computation in terms of the various hardware devices that constitute the computer system. A high-level description might describe the computation as the manipulation of some abstract data structure, like an array, which is a data structure that would normally not be implemented directly in hardware.

There are many high-level languages, some at a higher level than others. They support several different computational models. As an example, the language C++ supports a computational model usually called the object-oriented model. *Haskell* supports a paradigm called the functional model of computation. The problem with all high-level languages is that they support computational models which are not implemented directly by the computer hardware. Because of this, any program written in a high-level language, like C++, cannot be executed directly by a computer. In order for a program to be executable it must be written in a language that has a format that can be interpreted by the computer, and it must describe the desired computation in terms of the hardware available on the computer. This is what low-level languages do, and in particular, this is what is done by *machine language*, the lowest-level programming language.

1.1.2   Machine Language

Machine language describes a computation in terms of hardware, and, as such, is hardware dependent. That is to say that every processor model, for a general-purpose computer, has a different machine language, and these machine languages are non-portable. Another interesting fact about computers, for those used to interacting with them in a high-level language, is that computers only work with numbers. This is interesting because a program in C++ is written using words, like *if*, *void*, and *cout*, rather than just numbers. In fact, the user of C++ might well wonder what a numeric language would look like.

To illustrate what machine language looks like, let us examine an example. Consider the following C++ code fragment.

$$x = 5 + y * 3; \tag{1.1}$$

How might this appear in a numeric machine language? Consider the following machine code segment.

$$\begin{array}{l} \texttt{1, 1, 2, 3} \\ \texttt{14, 1, 1, 5} \end{array} \tag{1.2}$$

This segment might do the same computation as the C++ segment. To understand it, we must know something about the format of a machine language program.

The machine language we are using is fictitious, but similar to several actual machine languages. A machine language program is a sequence of instructions, written one per line. Each instruction, in this language, consists of four number, or *fields*. For our fictitious example, the fields have the following names.

op-code, destination, source, constant

Each instruction performs an operation on operands, and produces a result. Operations must be specified numerically in machine language, and so each possible operation the processor can perform is assigned a number to represent it, called its *op-code*. In the above example, the op-codes used are 1 for multiply, and 14 for add.

The op-code is followed by three operands. The operands in the C++ code are the variables *x* and *y*. Variables are a high-level concept. The corresponding concept at the hardware level is the register, which is a device that can store a number. A processor typically has a set of general-purpose registers available to the programmer, and they are often numbered. Each variable in the C++ program might be assigned a register, which is used in the machine language version of the code. In the above example, Register 1 (R1) is used to represent x, and R2 is used to represent y.

The instructions in Example 1.2 each specify two operands: the source operand, which is always a register, and a constant operand, which is always a constant value. The operation specified by the op-code is performed on these two operands, and the result is left in the destination register. If we were to write the above machine language fragment in a more human-readable form, we might produce something like the following.

$$
\begin{aligned}
R1 &= R2 * 3 \\
R1 &= R1 + 5
\end{aligned}
\tag{1.3}
$$

The first instruction multiplies R2 by 3, and stores the result in R1. The second instruction then adds 5 to R1, storing the result in R1.

We now can see the full scale of the problem of writing a program in a high-level language. In order to execute a C++ program, the C++ code must be rewritten, or translated, into this style of machine code.

1.1.3 Assembly Language

When the modern computer was first being developed, in the middle of the twentieth century, the only language available for programming a computer was machine language. As can be imagined, this was a difficult language for human programmers to use. Just reading through a program was difficult, requiring a mental translation between numbers and operations and operands. Mentally, this required the programmer to do what we did in the previous section, where we translated from numeric machine code to a notation that resembles assignment statements, which describes the semantics of the operations in symbolic form. These symbolic machine instructions are, essentially, assembly code. That is to say, *assembly language* is a symbolic form of machine code.

Assembly languages, like the one we used in the previous section, were developed to aid programmers of early computers. They are still useful today for several purposes, including as the target language of a compiler. The assembly example, Example 1.3, is, however, a little unusual in its notation. A more standard notation for the same code fragment might appear more as follows.

```
mult R1, R2, #3
add R1, R1, #5
```

This notation follows the format of the machine language instruction, starting with a symbolic op-code, a symbolic destination register, a symbolic source register, and ending with the constant.

1.2 COMPILERS AND ASSEMBLY LANGUAGE

The problem of translating from a high-level source language to a low-level machine language is a complex one. This is due to the fact that high-level languages typically have non-trivial syntax, and their semantics are much more involved than the semantics of machine language. As a consequence translation from a high-level language to machine language is typically done in stages, rather than all at once. First, the source code is translated to a middle-level language, which is often assembly language. In further stages the assembly language is translated into lower-level languages, eventually producing machine language as the result of the process. The idea is that splitting the translation process into stages spreads the complexity of the translation process over the stages, making each translation stage simpler.

1.2.1 Assembly Language Translation

In the first translation stage, a program called a *compiler* takes as input a source program in a high-level language like C++, and translates it into a program in a target language, which is at a lower level. The target language is typically assembly language.

We now have a reasonable idea of the form of both the source and target languages of a compiler. Although it is beyond the scope of this book to examine the translation done by the compiler in detail, it is worthwhile saying a little on the subject. A compiler inputs a source program and writes a new program in target code. The target program is the equivalent of the source program, meaning that the two programs do exactly the same thing. Given the same input, the target program will produce the same output as the source program.

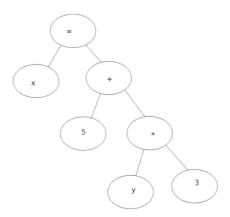

FIGURE 1.1 AST for Example 1.1.

1.2.2 The Translation Process

The translation of the source program is the result of a sequence of passes made through the source code. Two of the more important passes are *parsing* and *code generation*. Parsing is the task of deciphering the meaning of the statements of a program. This is done by building a structure out of the statements of the program, that is more machine readable than just simple text. As an example, a compiler might build what is called an *abstract syntax tree* (AST). For the example C++ code given in the previous section, Example 1.1, the AST might appear as in Figure 1.1. In this tree, intermediate nodes are operators, and leaf nodes are operands. The top intermediate node uses the assignment operator to assign a value computed by the right child to the variable *x*. The computation is calculated using an addition operator, and a multiplication operator.

Once the AST has been constructed, target code can be generated from the tree. This is done by traversing the AST, and at each intermediate node, outputting instructions that perform the specified operation. This code generation process, although complex, in its basics, is just printing out the contents of a tree, much as we might learn how to do in a course on data structures. The output of the code generator, however, follows a particular format; namely that the content is printed in the form of assembly instructions. Also, variable names must be replaced by their assigned registers, during the traversal.

1.3   THE ASSEMBLER AND OBJECT CODE

The compiler translates a high-level program into an assembly language program. The program must still be translated from assembly language to machine code. This is the job of another translation program called the assembler. However, technically, the assembler does not translate all of the way down to machine language, but rather into a format called *object code*. Object code can be thought of as incomplete machine code. A good analogy is a form, or boilerplate. A boilerplate is a complete document, except that it contains blanks that must be filled in with specific information. In the same way, object code can be thought of as machine code with blanks. The reason why such code would be produced, as opposed to complete machine code, is what we explain now.

1.3.1   External References

Larger programs in high level languages are typically split into modules. These modules are submitted to the compiler as separate files. So, for instance, you could have a C++ program composed of two modules: the module Q, and the module *Driver*. Splitting the program into modules, in this fashion, is an organizational aid for the programmer. Splitting the code also allows the programmer to work on one part of the code without touching other parts, and running the risk of inadvertently damaging them.

The compiler compiles only one module at a time. It produces assembly language versions of each module. These files are then submitted to the assembler, one at a time. The important point to remember in the following discussion is that the assembler sees only one module at a time, and while analyzing and translating, the module has no information about the other modules.

Modules may contain what are called *external references*. Consider the following lines that might be contained in the module Q.

```
extern int x;
x = 5;
```

The module Q is changing the value of a variable x. But the keyword *extern* declares that the variable is not in the module Q, but rather in another module. Continuing with the example, the module *Driver* might contain the following code.

```
int x ;
```

The assembly language versions of the two modules would contain instructions to mimic the C++ code. That is to say, the module Q would contain the following code.

$$\texttt{store x, \#5} \qquad (1.4)$$

This instruction writes the constant 5 to a memory location, designated as x.

The module *Driver* would contain a definition of the variable x. In assembly language, variables are typically stored in memory. A memory unit is an array of storage devices, called *words*. When referring to a particular word in the memory unit, in assembly language you would give that word a name, much as in the following definition that you might find in the module *Driver*.

```
x:    .word
```

This declaration allocates a word in memory for the variable, and associates the symbol x with it. This is equivalent to a variable declaration in C++.

In machine code, variables do not exist. When a program is executing, it is located in the computer's memory. As mentioned, memory consists of an array of *words*. In machine language, words are identified by their index. The indexes of the memory locations are called *addresses*. So, we might refer to the word at Address, or Location, 50, often written as M[50].

The instructions, as well as the variables of the program are stored in different words of the memory unit. In machine language, the variable x of our example, is simply a word in memory that has been allocated for the storage of data. That word is known only by its address. As a consequence, the *store* assembly instruction, in Example 1.4, might be translated into the following machine code.

$$\texttt{19, 50, 5} \qquad (1.5)$$

The assumption is that the *store* instruction has op-code 19, and that the variable x has been allocated the location M[50].

Here now is the problem with external references. In this example, the assembler attempts to write a machine instruction corresponding to the *store* assembly instruction. It knows what op-code to write, and it sees the constant value, 5, both from the assembly instruction. However it does not know the address of the variable x. The address of x is calculated when the module *Driver* is assembled, and since the assembler has no information about the module *Driver* when assembling the module Q, the assembler does not know the address of the variable x. Because of this, the assembler ends up writing out an instruction that looks something like the following.

```
19, x?, 5
```

This is an incomplete instruction. It contains a blank, indicated by the question mark, to be filled in later by some program that knows the location of the variable x. The blank contains a note indicating that the value to be filled in is the location of x.

These types of instructions, being generated by the assembler, are in fact the object code. As can be seen, these instructions can be thought of as machine instructions, but with blanks left in them. The reason that these blanks are required is mostly due to the inevitable presence of external references when programs are split into modules.

1.3.2 Compiler versus Assembler

As a final topic in this section, it is instructive to compare the assembler with the compiler. Both of these programs are translators. This is their similarity. There is, however, a significant difference between the two programs, in terms of the process they use to do the translation. The compiler's job is complex. The syntactic analysis, which ends up building the AST, is complex, and the code generation is also complex, often requiring that several assembly instructions be written for each node in the AST. The translation process for the assembler, however, is not all that complex.

Syntactic analysis is fairly simple for the assembler. An assembly instruction is, essentially, just a small sequence of symbols, as opposed to the highly structured form of high-level statements that the compiler deals with. And, once an assembly instruction has been decomposed into separate symbols, the code generation process is equally simple.

The process of writing out a machine instruction is referred to as *assembling* the instruction. This term captures the idea that the machine instruction is being built by connecting sub-parts of the instruction. The sub-parts of the instruction are derived from the symbols of the assembly instruction. Each symbol in the assembly instruction corresponds to a numeric value. For the example we give in Examples 1.4, and 1.5, the symbol *store* corresponds to the numeric value 19. The assembler could easily maintain the bindings between symbols and their numeric values in a table. To translate an assembly instruction into a machine instruction, then, simply requires that the assembler looks up the symbols of the assembly code in the table, procures their numeric values, and assembles the numeric values into machine instruction, following the format rules for the machine instruction.

Since the object code is incomplete, we still do not have an executable program. In order to fill in the blanks in the object code, all of the modules of the program must be examined simultaneously. This, then requires another step in our translation process.

1.4 THE LINKER AND EXECUTABLE CODE

The linker is a program that inputs a set of object code files representing a program, and outputs what is called *executable code*. Executable code, usually, is complete machine code. The executable code is written to a mass storage unit such as disk or flash memory, as a file. In order to do this, the linker must resolve all unresolved external references. Including this job, the linker has three primary jobs.

- Resolving external references.

- Library searches.

- Relocation of module code.

We start our discussion of the linker with the resolution of external references.

1.4.1 Resolving External References

The input to the linker consists of all of the object files for the whole program. For our example, the linker would examine both the object file for the module Q, as well as the object file for the module *Driver*. The linker would notice that the module Q contains a blank to be filled in with the address of the variable x. It would then examine the module *Driver*, and determine the address of x. The blank in the instruction in Module Q would then be replaced with the correct address. This is a basic description of the procedure used by the linker to perform its most important job: resolution of external references.

1.4.2 Searching Libraries

The linker's second job is performing library searches. Suppose that the module *Driver* contains the following C++ code fragment.

$$z = \text{sqrt(y)}; \qquad (1.6)$$

What this code does is to call a function *sqrt*, pass it an argument y, and place the return value into the variable z.

To implement the function call of Example 1.6 in machine language, a method for passing arguments to a function, and passing a return value back from the function, must be agreed upon. Every processor supports some such method, called the *calling conventions* of the processor. For our simple processor, we have assumed that arguments are passed to the function by loading them into special-purpose *argument registers*, whose names begin with the letter A, as opposed to the letter R, to distinguish them from the general-purpose registers of the processor. To access the arguments, the function simply accesses the argument registers. When the function is ready to return a value, it loads its return value into another special-purpose register, called the *return value* (RV) register. This register is then accessed by the calling program, to receive the return value. The registers associated with the calling conventions are called *special-purpose registers*, because the programmer would only use them for function calls, as opposed to general-purpose registers whose use is unrestricted.

When translated into assembly language, the sequence from Example 1.6 might resemble the following.

$$
\begin{array}{l}
\texttt{load A0, y} \\
\texttt{call sqrt} \\
\texttt{store z, RV}
\end{array} \tag{1.7}
$$

In Example 1.7, the assembly fragment first fetches the operand y from memory, and loads it into the argument register A0. Then a *call* instruction is executed, which causes the processor to jump to the address *sqrt*. When the *sqrt* function executes a *return* instruction, the processor jumps back to the *store* instruction following the call. At this point the return value is transferred from the RV register into the memory location at address z.

The call to the function *sqrt* constitutes an external reference. Clearly *sqrt* is not defined in the module *Driver*. But, it is also not defined in the module Q. Most C++ programmers know that the function *sqrt* is, in fact, defined in the C++ library, which is a collection of modules, in object code format. *Sqrt* is defined in the module *math*. In order to allow a program to use code contained in library modules, we must alter our procedure for external reference resolution to include library search. That is to say, when the linker encounters an unresolved reference, it first looks for resolution in the actual modules of the program. If the linker does not find a definition for the reference, then it searches all of the modules in all of the libraries it can find. If the linker finds a definition for the reference in a library module, that module is added to the other modules of the program. If the reference is still unresolved after this rather exhaustive search, a *linker error* occurs, and linking is aborted.

1.4.3 Relocation

We now turn our attention to the last job of the linker: relocation of modules. In the beginning of our discussion of the linker, we said that when resolving the external reference to the variable x, the address of the variable is calculated by examining the module *Driver*, which contains the variables definition. It was, however, not explain clearly how addresses are calculated.

Addresses inside a module are relatively easy to calculate. The module has a base, or beginning, address. The objects in the module are either instructions or data allocation declarations. The number of words taken by each instruction is known, and the data declaration explicitly provides the number of words being allocated. It is easy to run through a module from top to bottom, and track the number of words written out at any particular point in the module code. In this way the address of any object can be calculated by using the current word count as an offset, which is added to the base address. This job is readily performed by the assembler.

The situation is complicated by the fact that our program is not a single module. However, when the linker finishes its work, it must produce a single monolithic program, occupying a solid, contiguous block of memory. In other words, it must combine the different modules into one block. To do this, the linker must order the modules in the memory block. For our example, the linker might decide to place the module *Driver* first in the machine language program, and then follow it with the module *Q*.

In term of addressing, the module *Driver* would be assigned the base address 0. Let us assume that *Driver* contains 3,000 words of machine code, and that *Q* contains 2,000 words. Then the base address of *Q* would be 3,000, and it would end at address 4,999, which is one subtracted from the sum of 3,000 and 2,000. The upshot of this calculation is that, if each module has been assembled, assuming a base address of 0, all addresses in the module *Q* must be adjusted by adding 3,000 to them. This process of adjusting addresses, after deciding on module order, is the process of module relocation.

1.5 THE LOADER

As discussed, the linker produces executable code, that is saved to mass storage. Although the code produced is executable, it is not yet running. The program that starts the executable code running is the loader. What does it take to start a program running? Firstly, the program must be loaded into memory from mass storage. This process is often termed *program relocation*. Notice that the linker and the loader both do relocation. The difference is that the linker is relocating modules, or pieces of a program. The loader, on the other hand, relocates whole programs in memory.

1.5.1   Processes and Workspaces

The modern computer is typically a *multi-process* system. This means that the operating system is often running several processes, or programs, simultaneously. Technically, this is impossible, since the processor can only do one thing at a time. Technically, the processor is only making it appear that it is running several programs simultaneously. In reality it is sharing time between the different running processes, working one process for a small amount of time, then switching to another process for a small amount of time, then on to another process, and continuing this procedure, executing small pieces of code from each program. Because the amount of time spent on each program, called the *quantum*, is so small, and the processor returns to the same process in such a short time, it appears to a human observing a particular process that the processor is working on that process without interruption. To the human observer, it appears that the processor is working on several processes, each one uninterrupted, although this is not actually true. Perhaps a more accurate statement would be that, in the modern processor, several processes are active at the same time.

Each process that is active is assigned a part of memory to work with, often referred to as the process's *workspace*. The workspace is allocated to the process by the operating system when the process is started.

Our discussion of the process workspace is leading into an explanation of why loading a program into memory requires relocation. Let us reconsider our continuing example. We wish to run the *Driver* program. It has been compiled, assembled, and linked, and is now in the form of an executable file on disk. To run it, the operating system starts a new process, and assigns a block of memory, its workspace, to the process. Now the loader reads the executable file from disk, and writes the code to the workspace. The problem that the loader has, is that the workspace may be located anywhere in memory. This is a problem because when the linker previously wrote out the executable code, it did not know what the base address of the program would be. What it did is the equivalent of writing code that starts at memory location 0. This is almost certainly not going to be the case.

When the loader loads the program into the workspace, if there is any hope of the program running correctly, all addresses in the program must be altered from using a base address of 0 to using the base address of the workspace. There are several ways of performing the alteration. Often the alteration does not require an explicit change to every address in the program. Often the changes can be made simply by initializing a special-purpose base address register to the base address of the workspace. Whatever the method used, the point is that somehow the executable code must be made to execute at the location where it is loaded into the workspace. This process, which we have already mentioned is called relocation, is performed by the loader.

1.5.2 Initializing Registers

So, now the *Driver* program is loaded into the workspace. The next job of the loader is to initialize the processor in preparation for executing the program. Many of the registers in the processor must be initialized. This is truer of the registers that are used by the processor for specific purposes. In particular, one such register in the processor is often called the *program counter* (PC). The job of the PC is to keep track of where in the program the processor is currently executing. The PC always contains the address in memory of the next instruction to be executed. Every time the processor finishes executing an instruction, it fetches the instruction indicated by the PC from memory, and begins executing that instruction. The PC is also updated to point to a new, next instruction, usually by incrementing it.

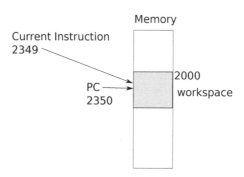

FIGURE 1.2 The PC and memory.

The role of the PC is illustrated in Figure 1.2. The diagram shows the process workspace in memory. The base address of the workspace is 2000. The PC contains the address 2350. The instruction currently being executed would then be at address 2349. When this instruction has been completed, the machine instruction at 2350 would be loaded into the processor to be executed, and the PC would be incremented to 2351.

It should be clear that the PC must be initialized to the base address of the workspace, when the program begins execution. This then is one example of several special-purpose registers which the loader must correctly initialize before the program begins execution. Actually, the initialization of the PC is done as the last register initialization. The reason for this is that if the loader changes the PC to point to the starting instruction of the *Driver* program, the next instruction executed will be the first instruction in the *Driver* program. In other words, the loader has jumped to the start of our program, and started it up. This is the last action of the loader; after loading the program, and initializing registers, it starts the program running.

1.6 SUMMARY OF THE TRANSLATION PROCESS

We now have some sort of understanding of the answer to our original question as to how a C++ program is executed. The process of getting the program to run is actually fairly sophisticated, and involves several steps. These steps are summarized in Figure 1.3.

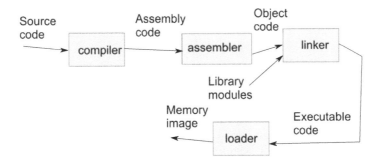

FIGURE 1.3 Workstream for source code translation and execution.

Figure 1.3 shows the compiler first translating the source code program into assembly language. The assembler then translates the program down into object code. The linker links the object modules together, and adds in any required library object modules. The output of the linker is executable code, which is loaded into memory by the loader.

1.7 THE PROCESSOR

We have now partially answered the question as to how a C++ program executes on a computer. We have examined the translation process that produces and starts machine code. But if we dig deeper, we discover that this is only part of the answer. We know how to get the program to run, but not how the processor actual executes the instructions. To answer that part of the question, we must examine what sort of machine the processor is, and how it is designed. This is the point at which we must start looking at the circuitry in the processor. We stop talking about the software used, like the compiler and linker, and start talking about the hardware device from which the processor is built.

When we discussed the software tools used to translate a program, we did so by describing language levels. We use this same level approach when discussing hardware. We have two major levels of abstraction used when specifying processor design.

- The *register transfer level* (RTL), or *behavioral level*.

- The *gate level*, or *structural level*.

The RTL level is the higher level of abstraction, and the gate level is the lower level. As with the software, we will be examining the two levels from the top down, starting with the RTL level.

1.7.1 Processor Behavior

To begin our discussion of the RTL level, we examine the functioning of the processor. A processor is a device that fetches instructions stored in memory, and executes them. This is a slight simplification of the functioning of the processor, but not much of one. The processor performs a sequence of steps, over and over, as long as it is supplied with electrical power. This sequence of steps is called the *machine cycle*, or *instruction cycle*. The instruction cycle is a three-step procedure.

1. Fetch an instruction from memory.

2. Decode the instruction.

3. Execute the instruction.

In Step 1 the processor reads the next instruction to be executed, as indicated by the PC register, and brings it into the processor. The processor also increments the PC at this time. In Step 2 the processor splits the instruction into fields. By so doing it discovers what operation is being performed, and on what operands.

In Step 3 the appropriate circuitry in the processor is activated to execute the operation on the operands. The result of the operation is then written back to the destination operand. The cycle is then repeated, to execute the next instruction.

1.7.2 Processor Structure

We now have described what the processor does. Let us now look at the structure of the processor. The processor is a machine composed of several devices. For instance, we have already discussed the fact that the processor contains a set of both general-purpose and special-purpose registers. These registers, collectively, form a device called the *register file*. Another device called the *arithmetic-logic unit* (ALU) performers arithmetic and logic machine operations.

All of the devices in the processor must be connected, in order to communicate with each other. How the devices are connected is called the *data-path* of the processor. The different devices must also be told when to perform their function and often which of several functions to perform. The circuit that controls the devices, in this fashion, is called the *control unit*.

1.7.2.1 The Data Path, Registers, and Computational Units

The best way to explain a data path is with an example. We will be building an example that contains just two registers. The circuit will perform just two operations, as follows.

$$R1 \leftarrow R1 + R2$$
$$R2 \leftarrow 0 \tag{1.8}$$

In the first operation of Example 1.8, the contents of the registers R1 and R2 are added, and the result is placed into the register R1. In the second operation, the register R0 is set to 0.

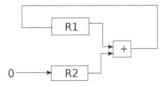

FIGURE 1.4 Example data path for Example 1.8.

The data path of this device is the circuitry that allows these two operations to be performed. This data path would include two major components: registers, and for each register, a *computational unit*. A computational unit computes the new value of a register, resulting from an operation. In our example, we would have one computational unit to compute the new value of R1, as the sum of R1 and R2, and another that computes the new value of R2 when it is cleared. All computational units and registers must be connected in a fashion so as to make every operation possible. Figure 1.4 shows how a circuit might be connected for Example 1.8.

Figure 1.4 shows two registers: R1 and R2. An adder computes the sum of the two registers. The two operations that we must compute can be performed as follows. The operation $R1 \leftarrow R1 + R2$ is performed by allowing the adder to compute R1 + R2. The result of the adder is connected as the input to R1, thus becoming the new value of R1. The operation $R2 \leftarrow 0$ is performed with the number 0 connected to the input of R2, thus becoming the new value of R2.

The circuitry in Figure 1.4 can be split into two types of devices: registers and computational units. The registers are, obviously, R1 and R2. The inputs of the registers are connected to computational units. These are devices that compute the next value of the register, after an operation is performed. In our example, the computational units are rather simple. For R1, the computational unit is simply the adder, and for R2, the unit is simply the hard-wired value of 0.

The circuit in Figure 1.4 is capable of performing the required operations. When the operations are performed is another matter. The diagram leaves this question open. The two operations might be performed separately, simultaneously, or even not at all. Determining under what conditions the different operations are performed is the job of the control circuitry.

1.7.2.2 Control Circuitry

Continuing with our example, we must now discuss our intent for the circuit. Suppose that we have two wires, S_1 and S_2. We want the operation with R1 to be performed when we send an electrical signal on the wire S_1, and we want the operation on R2 to be performed when we send a signal on the wire S_2. We might write these control specification as follows.

$$S_1 : R1 \leftarrow R1 + R2$$
$$S_2 : R2 \leftarrow 0$$
(1.9)

Let's assume that each register is capable of two possible operations: *lock*, in which the register keeps whatever value it has, and *load*, in which whatever is on the input port of the register becomes the new register value. The registers must be told which operation to perform, and so we add an extra input line to the registers, called *load* (LD). This line normally has no signal on it, indicating that the register is locked, and should hold its value. When the LD line is given a signal, a load operation is specified, and the register should input a new value.

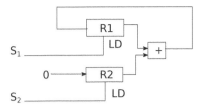

FIGURE 1.5 Example control circuitry for Example 1.9.

With these assumptions we can redraw our circuitry figure to include the control inputs, S_1 and S_2. In Figure 1.5, the lines S_1 and S_2 are connected to the load lines of the registers. When S_1 is activated, the register R1 is opened, and receives the results of the adder's computation, as desired. Also, as desired, when the line S_2 is triggered, the register R2 is opened, and receives the 0 value.

Our new diagram now has three different types of circuits in it. It contains registers and computational units, forming a data path and control circuitry. The control circuitry is the wiring that causes the specified operations to happen, when desired. For our example, the control is simple, and consists of just the wires S_1 and S_2.

The example we have given is a description of the RTL (register transfer language) level of a very simple circuit. But this type of description can, in fact, be developed for a full-fledged processor. The description would define the operation of the processor, in a precise fashion, allowing the description to be implemented as actual hardware.

1.8 DIGITAL CIRCUITRY

At the RTL level, the processor is a collection of connected devices. These devices communicate with each other to perform operations. So, we now know something about what type of a machine the processor is. But, our answer to what type of a machine a processor is, only succeeds in raising several other questions. In particular, we have the concept of a device, such as a register or an adder, but then we might start wondering how these devices are built. This leads us to examine the level of circuitry below the RTL level.

Below the RTL level is a level which is sometimes called *gate-level circuitry*. It is called gate level because these circuits are constructed out of components that are called logical *gates*. A gate is a component that computes a Boolean

FIGURE 1.6 The AND gate.

value. As an example, Figure 1.6 is the schematic representation of an AND gate. This gate computes the Boolean function $z = a \wedge b$. The inputs to this gate, a and b, and the output, z, are Boolean signals. Of course, the signals sent to the gate are electric voltages, not values of true or false. But the voltages sent to the gate are interpreted as Boolean values. For instance, a voltage of 5V might be interpreted as a *true* value, and a voltage of 0V might be interpreted as a *false* value.

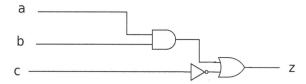

FIGURE 1.7 Combining gates into larger circuits.

Gates can be connected into larger circuits, that compute more complex Boolean functions. For example, the schematic in Figure 1.7 shows a circuit that computes $y = a \wedge b \vee \neg c$. The diagram uses an AND gate, connected to a and b, an inverter, or NOT gate, with c as input, to compute $\neg c$, and an OR gate, which produces the output z.

The circuits described at the gate level can become much larger and more complex. All of the devices used in our RTL-level examples, Examples 1.8 and 1.9, can be described as collections of connected gates. The adder will be designed later in this book. The registers can also be described as collections of gates. Later, we will design registers, using a component called a *flip-flop*. Flip-flops are devices that can store Boolean values. Although we treat them as atomic, they are actually constructed from gates that are similar to the AND gate and the inverter.

1.9 SUMMARY

In this section we have examined how a high-level source program is executed. We worked our way from the source code, and saw how it was translated, first into assembly language, next to object code, and how it was finally loaded into memory as a working machine language program.

We then examined the hardware that is used to execute the program. We saw that it can be viewed from several levels. At a high level, it is a collection of communicating devices, performing RTL-level operations. Underneath is a circuit built out of simple components that compute Boolean functions.

Our examination of the processor has so far been simplified, and only cursory. In the remainder of the book we revisit each of these levels in more detail. We start with the low levels, discussing hardware. In Chapters 2, 3, 4, and 5 we work our way up from the gate level to the RTL level. Then we examine the machine language level in Chapter 6. Once we have a good grasp of architecture issues, and a grasp of RTL design, we then use this information to examine processor design in Chapters 7, 8, and 9. We finish with Chapter 10 which covers, briefly, some more advanced topics.

1.10 EXERCISES

1.1 Draw the AST for the following C++ statements.

    a. m = (a + y) * (x - 3);

    b. n = a * b * 5;

    c. p = y + v / 4;

1.2 Compile the C++ statements from Exercise 1.1 into assembly code. Use the op-codes *div* and *sub* for division and subtraction, respectively.

1.3 Assemble the following assembly code into machine code. Assume that the machine language op-codes for *load*, *store*, *mult*, *add*, *div*, and *sub* are 18, 19, 13, 14, 15, and 16, respectively. Also assume that the variable x is stored at location M[50].

    a.
```
load R1, x
mult R2, R1, #9
store x, R2
```

b. `sub R0, R1, #8`
   `div R2, R0, #2`

1.4 A security issue exists in many multi-process systems; one program may use an address in another program's workspace. This should not be allowed. Explain a way of preventing this. Discuss whether your solution requires runtime components, load-time components, assembly time components, all of these, or components used at other than these times.

1.5 Draw a diagram showing both the data path and control circuitry for the following set of operations. Notice that V_2 causes two operations to be performed simultaneously.

$$
\begin{aligned}
V_0 &: R1 \leftarrow R1 + 3 \\
V_1 &: R2 \leftarrow R1 \\
V_2 &: R3 \leftarrow R2, R4 \leftarrow 6
\end{aligned}
\qquad \text{(P1.1)}
$$

1.6 Draw gate-level circuits to implement the following Boolean functions.

a. $v = (a \vee b) \wedge c$

b. $w = \neg(a \wedge b)$

c. $x = (a \wedge b) \vee (\neg a \wedge \neg b)$

Number and Logic Systems

CONTENTS

In order to understand the workings of a computer processor, you will need to be familiar with the mathematics upon which computer systems are based. In this section, we examine the background you will need to understand the rest of the material in the book.

When we start trying to represent numbers on a computer, we use the *binary number system*. We use binary as a result of the type of circuitry we employ in computer construction. And, while binary is a natural way of representing numbers on a computer, it is not a very natural system for human use. To make binary more accessible to the human user, we use the *hexadecimal number system*. This number system is a little easier for people to use, and it is still relatively easy to convert hexadecimal representations into binary. We cover both the binary and hexadecimal number systems in the first part of this chapter.

Computer circuitry is largely an application of *Boolean algebra*. In the second part of this chapter we cover the basics of Boolean algebra.

2.1 NUMBERS

Integers are typically represented in computer systems, in binary or base-two. Binary is a number system with just two digits: 0 and 1. A binary digit is is often referred to as a *bit*.

We store an integer number as a collection of bits, usually of fixed width. This width is the width of the storage unit for an integer, called the processor *word*. So, for example, if the processor word is eight bits, then the number 13 (1101 in binary) would be represented as 00001101. But, an interesting question is why the original bit string, 1101, is, in fact, the decimal number 13.

Let us examine this question by starting with the binary number 00110101, and ask what decimal number this represents. Both binary and decimal are *positional* number systems. This means, for instance, in the decimal number 365, the fact that the digit 3 is in the leftmost position, means something different than it does in the decimal constant 653, where it is in the rightmost position. In particular, in the decimal constant 365, each digit is multiplied by a distinct power of 10. That is, $365 = 3 \times 10^2 + 6 \times 10^1 + 5 \times 10^0$. This series is called the *decimal expansion* of 365. We see that digits to the left are multiplied by larger powers of ten than digits to the right, and so we refer to digits to the left as *high-order* digits, and those to the right as *low-order* digits.

The decimal system gets its name from the fact that, when computing the powers, the *base*, or *radix* used, is 10. In fact, the decimal notation is often referred to as base 10. In base 10, there are 10 digits, 0 through 9, and positions represent powers of 10, starting at the right position, representing 10^0, and working left, with increasing exponents.

TABLE 2.1   Successive division.

Calculation	Quotient	Remainder
$365 \div 2$	182	1
$182 \div 2$	91	0
$91 \div 2$	45	1
$45 \div 2$	22	1
$22 \div 2$	11	0
$11 \div 2$	5	1
$5 \div 2$	2	1
$2 \div 2$	1	0
$1 \div 2$	0	1

Binary is base 2. This means there are two digits, 0 and 1, and the different positions represent different powers of 2. Returning to our example, 00110101 can be expanded using its *binary expansion* as $00110101 = 0 \times 2^7 + 0 \times 2^6 + 1 \times 2^5 + 1 \times 2^4 + 0 \times 2^3 + 1 \times 2^2 + 0 \times 2^1 + 1 \times 2^0$, or more succinctly as $00110101 = 2^5 + 2^4 + 2^2 + 2^0$, eliminating all zero terms, and all multiplications by one.

Binary expansion gives us a tool for converting any binary number into decimal. We simply expand the binary number, and do the arithmetic in decimal, obtaining a decimal result. For our example, $00110101 = 2^5 + 2^4 + 2^2 + 2^0 = 32 + 16 + 4 + 1 = 53$.

With a systematic way of converting any binary number into decimal, we soon realize that we do not, as of yet, have a way of converting a decimal number into binary. To do that job, we use a method called *successive division*. This method uses a property of integer division, when the dividend is the radix. To start with, let us do an example in decimal to demonstrate this property.

Let us examine the number 365. We compute $365 \div 10$. We get a quotient, 36, and a remainder, 5. What we have discovered is that by dividing by the base, 10, we can pull off the low-order digit of a decimal number, which is the remainder, and separate out the remaining digits, which constitute the quotient. Continuing by computing $36 \div 10$, the quotient divided by 10, we get a new quotient of 3, and a remainder of 6. We now have the second-lowest digit of 365. We finish by computing $3 \div 10$, yielding a new quotient of 0, and a remainder of 3. The 3 is the last digit of 365. At this point we have all of the decimal digits of the number 365.

We now apply this method to a decimal number, but instead of dividing by 10, we divide by 2. Note that this will pull off binary digits (0 or 1 will be the remainder), rather than decimal digits. We use 365 as our example, successively dividing our quotient by 2. The process is shown in Table 2.1. We stop with a quotient of 0, since any digits we pull off after this will be 0.

To construct a binary number out of this calculation, we observe that, examining our decimal example, the first division pulls off the rightmost digit, and successive divisions pull off digits further to the left. So, to construct the binary equivalent of 365, we write the remainders in Table 2.1 from top to bottom, right to left. Doing this gives us the binary number 101101101.

We now have conversion methods allowing us to move freely between decimal and binary. Binary and decimal are the two most interesting number systems, in terms of computer design; binary, because digital circuitry works in binary, and decimal, because we humans, predominantly, use decimal. A third numbering system, hexadecimal, or base 16, is also important, in terms of computer design. It is important because of the clumsiness inherent in the use of binary. That is to say that binary numbers are not at all compact. For instance, we just saw that the decimal number 53 has the long binary representation of 110101. It would be nice if we had a number system which was easily converted into binary, but much more compact, like decimal. Hexadecimal is one such number system.

2.1.1 Hexadecimal Numbers

Hexadecimal, as previously stated, is base 16. That means that numbers are represented using 16 digits, and the powers in the hexadecimal expansion will be powers of 16. Because we only have 10 decimal digits, 0 through 9, when writing hexadecimal numbers we normally augment this set with six more digits, represented by the letters A through F. The digit A represents a value of 10, digit B represents the value 11, digit C represents the value 12, digit D represents the value 13, digit E represents the value 14, and digit F represents the value 15. So, for example, the hexadecimal number A3F represents $10 \times 16^2 + 3 \times 16^1 + 15 \times 16^0$.

An interesting property of hexadecimal numbers is their relationship to binary numbers. The property we are referring to is that four binary digits are represented by one hexadecimal digit. Using the above example of the hexadecimal number A3F, if we were to write a 4-bit binary representation of 10, 1010, for the A, a 4-bit binary representation of the 3, 0011, and a 4-bit representation of 15, 1111, for the F, and then we were to concatenate them into one big bit string, 1010 0011 1111, this would, in fact, be the binary representation of the hex number A3F.

A reverse process can be used to convert from binary to hexadecimal. For example, the binary number 0010111010001011 could be split into groups of

4 bits, 0010 1110 1000 1011, and then each group replaced by its corresponding hex digit, yielding the hexadecimal representation of this binary number: 2E8B.

2.1.2 Adding Binary Numbers

We could easily say that what makes numbers different from just a collection of digits are the arithmetic operations we perform on them. Perhaps the most basic operation we perform on numbers is the arithmetic operation of addition.

If you look at an addition problem performed in decimal, you might do the calculations in a way somewhat similar to the following example.

$$
\begin{array}{r}
\scriptstyle 01100 \\
0365 \\
+1192 \\
\hline
2557
\end{array}
\qquad (2.1)
$$

In this long addition process you add the addends to each other, column by column. In each column you have a *carry-in* value, coming from the previous column, and in each column you produce a sum digit, written at the bottom of the column, and a *carry-out* digit, which becomes the carry-in to the next higher column.

Example 2.1 is similar to what you would do to add two decimal numbers, but probably not exactly what you would do. We have made a couple of modifications to the calculation, to help us understand how binary addition is carried out on a computer. The first modification is that we have used fixed-length numbers, of four digits. In a normal addition calculation, you would most likely add numbers of varying lengths, and eliminate the leading zero in 0365. The second modification we have made is to rigorously show all carry-in and carry-out values for each column. We show a default carry-in value of zero for the low-order column, and a carry-out from the high-order column, also, of zero.

When adding in binary, we use the same method used in decimal, but modified to the binary system.

- In decimal we obtain a non-zero carry-out in a column, if the result of adding that column is at least ten. In binary we obtain a non-zero carry-out if the result is at least two.

- In decimal we obtain the sum digit at the bottom of the column, when there is a non-zero carry-out, by subtracting ten from the column sum. In binary we subtract two.

In both decimal and binary, the carry-out will always be either 1 or 0. These changes are illustrated in the following binary example.

$$
\begin{array}{r}
{\scriptstyle 00110} \\
1011 \\
+0011 \\
\hline
1110
\end{array}
\qquad (2.2)
$$

You can examine Example 2.2, and verify that we have used the long division method described above to calculate the sum and carry-out bits in each column. For example, in the low-order column, the carry-in is 0, and the two addend bits are 1, resulting in $0 + 1 + 1 = 10$ (a result of 2 in decimal). Since the result is at least two, we subtract two from it, giving us a sum bit of 0, and a carry-out from this column of 1.

2.1.3   Representing Negative Integers

So far, we have not yet talked about negative numbers. In our representation of numbers on a computer, we have described a process by which the bit configuration in a word is converted to decimal by taking the binary expansion of the bits in the word. If you do this, the result you get will always be non-negative.

Numbers that are always non-negative are typically called *unsigned* integers. We need a representation that will be able to cope with *signed* integers. That is to say, we need a representation that can store a number with a sign, either positive or negative.

The overall strategy, with signed integers, is to reserve half of the bit configurations for negative numbers, and half of them for non-negative numbers. A simple way of accomplishing this is to use the high-order bit of the word as an indicator of the sign of the number. This bit is referred to as the *sign bit*. In most applications, the way it is used is as follows.

- If the sign bit is 0, then the number is non-negative.

- If the sign bit is 1, then the number is negative.

For example, the 8-bit configuration 00101110 would represent a positive number, since the high-order bit, or sign bit, is 0. (This number is actually 46 decimal.) The bit configuration 11101001 would be interpreted as a negative number, since the sign bit is 1. Historically, several ways of representing negative numbers have been used. Initially, you might decide to represent numbers the same way we humans do. So, for example, we write a number like −623 using a notation consisting of a sign, in this case a negative sign, and a magnitude, in this case 623. Similarly, we write positive numbers, like +745, giving

TABLE 2.2

Notation	107	−107
Sign-magnitude	0 1101011	1 1101011
One's Compliment	0 1101011	1 0010100
Two's Compliment	0 1101011	1 0010101

the sign, and the magnitude, although quite frequently we just leave off the positive sign.

This notation, in which we give the sign and magnitude, is called *sign-magnitude notation*. You might wonder how sign-magnitude notation works in binary, so we will present a couple of examples. Let's take two numbers: 107, and −107. We show how these two numbers are represented in an 8-bit format, using sign-magnitude. In the first row of Table 2.2 we give the representations for sign-magnitude notation. The number 107 is simply converted into the binary representation, 1101011, and a positive sign bit is tacked onto the high-order bit position. For the number −107, we use the same binary magnitude, and tack on a negative sign bit.

Although sign-magnitude seems like a reasonable way of including representation for negative numbers, it turns out to present a few problems. In particular, one major problem is that there are two representations for zero: 00000000, often referred to as positive zero, and 1000000, often referred to as negative zero. This may seem a rather minor problem, but it is, actually, immense.[1] And so, the sign magnitude representation for integers, used in some older computer systems, has mostly disappeared.

A second method of representing signed integers is *one's compliment notation*. In one's compliment, we simply invert the bits of a positive number to obtain its negative. That is to say, we replace all 0 bits with 1 bits, and all 1 bits with 0 bits. We show the one's compliment notations for 107, and −107 in Table 2.2, on the second row. The number +107 is simply the 8-bit binary representation of 107. For the representation of −107, we have inverted each of the bits in the magnitude, 107.

Notice that, in this representation, the sign bit is handled correctly; it is 0 for non-negative numbers, and 1 for negative numbers. But, notice also that

---

[1] One of the more common operations a program will perform is to check if $a = b$, where a and b are two integers. This is actually checked by calculating the value of $a - b = 0$. When you have two values of zero, you are never sure which value will be produced by the subtraction $a - b$. If you are unlucky, the result will be −0, which will then be compared to +0. The result is that the test will fail, when it should succeed, causing the program to behave in what seems to be an inexplicable manner.

this notation has the same problem with zero; there is a –0, 11111111, and a +0, 00000000. So, this leaves us looking for a notation that solves this zero problem.

The system almost universally in use today is *two's compliment*. In this system, the negative of a number is calculated by taking its two's compliment. Let us examine how you would calculate the two's compliment of a number. As an example we will perform the calculation on +107 = 01101011.

The two's compliment calculation is performed in two steps

1. Take the one's compliment. For our example this would be 10010100.

2. Add one to the result of the previous step. For our example, this yields 10010100 + 1 = 10010101. This is the two's compliment.

Table 2.2 shows the two's compliment representation for our examples: +107, and –107. Notice that, as with the other two notations, two's compliment also handles the sign bit correctly.

The two's compliment notation turns out to be a fairly well-behaved way of representing integers, in terms of arithmetic. Firstly, there is only one value for zero. To see this, let us try taking the two's compliment of 0, to get –0. This calculation is shown below.

$$
\begin{array}{r}
\scriptstyle 111111110 \\
11111111 \\
+1 \\
\hline
00000000
\end{array}
\tag{2.3}
$$

Calculation 2.3 shows what happens when we add one to the one's compliment of 0. As usually happens in addition, the carry-out is discarded, and the result we get is 00000000. In other words, the two's compliment of 0 is just 0 again.

A second desirable property, for any representation of negative numbers, is that, for an integer a, $-(-a) = a$. To demonstrate that this is true, consider the binary number 00011010 (26 in decimal). If you take the one's compliment of this, 11100101, and add one to form the two's compliment, you get 11100110 (–26 in decimal). Now if you use this number, 11100110 (–26 in decimal), take its one's compliment, 00011001, and add one to this, producing the two's compliment of –26, you get 00011010 (26 in decimal).

A third property that is desirable is that for any integer a, $a + -a = 0$. We say that a and $-a$ are *additive inverses* of each other. We demonstrate the additive inverse identity by adding 26, and –26.

$$\begin{array}{r} \scriptstyle 111111100 \\ 00011010 \\ +11100110 \\ \hline 00000000 \end{array}$$

(2.4)

There is an easier way to compute the two's compliment of a binary number. It is based on copying, and slightly transforming, sub-strings of the binary string. This method is best illustrated with an example.

If you take the string 11001100 (decimal –52), you can split it into three sub-strings, which we list, from right to left.

1. *Trailing zeros.* The string of zeros on the right of the string, up to the first occurrence of 1. For our example, the trailing zeros are the substring 00. For some strings, this group may be absent. This sub-string is copied to the result, as is.

2. *First one.* The rightmost occurrence of a 1, that marks the left end of the trailing zeros. For the string representing 0, this group may be absent. This sub-string is copied to the result, as is.

3. *The rest.* All bits left of the first one. In our example this would be the substring 11001. For the string representing 0, this group would be absent. The one's compliment of this sub-string is copied to the result.

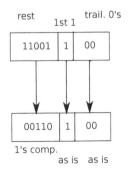

FIGURE 2.1 Copy method to calculate two's compliment.

With each step, we indicate how the substring is copied to form the result. Figure 2.1 shows how this copy shortcut is performed for –52. Writing –52 with the three groups, *the rest, first one,* and *trailing zeros,* in that order, we get 11001 1 00. We copy *the rest* after inversion, as 00110, the *first one* as 1, and the *trailing zeros* as 00, yielding 00110 1 00, which gives the result 00110100 (52 decimal).

With two's compliment notation, we have a good way of representing signed numbers, at least in terms of algebraic properties. Just like sign-magnitude, and one's compliment notations, a 1 in the sign bit indicates a negative number, and a 0 in the sign bit indicates a non-negative number. Half of the bit configurations represent negative numbers, and half represent non-negative numbers.

2.2 BOOLEAN ALGEBRA

We now turn our attention to a tool used in circuit design. Computer circuitry is based on Boolean algebra. Boolean algebra is a mathematical algebra with two values: *True* and *False*. In our discussion, we will be using the value 0, to represent *False*, and 1, to represent the value *True*. This is the usual notation used when Boolean algebra is used in circuitry.

2.2.1 Boolean Functions

Boolean algebra is a mathematical algebra, similar to other algebras, like arithmetic algebra. Algebras consist of a set of values, and a collection of operators that can be applied to the values. Using the operators we form expressions that manipulate values. An example using arithmetic algebra follows.

$$x + 2 \cdot y \tag{2.5}$$

In Expression 2.5 we see values, like the constant 2, and variables, like the variables x and y, being operated on by operators, like addition and multiplication. The operators act on the values of the algebra. Variables are used when we do not know the value of an operand, and so we leave its exact value open.

Boolean algebra consists of the Boolean values, *True* and *False*, and a collection of operators that operate on the Boolean values. From the operators and Boolean values we can form expressions describing a sequence of operations. The following expression is an example of a Boolean expression.

$$a \cdot b + \overline{a} \cdot \overline{b} \tag{2.6}$$

In Expression 2.6, although there are no explicit Boolean values, we have variables, a and b, that can take either of the values *True* or *False*. We also have three different operators: the "+" operator, which is a binary operator, called OR, the "·" operator, called AND, and the over-line operator, called NOT. What these operators do to their Boolean operands is what we are now going to discuss.

Boolean algebra has three basic operations. We can describe what they do to their operands using a *truth table*. This is a common method of specifying

TABLE 2.3   Truth table for
AND, and OR.

a	b	$a \cdot b$	$a + b$
0	0	0	0
0	1	0	1
1	0	0	1
1	1	1	1

TABLE 2.4   Truth table
for NOT.

a	$\overline{a}$
0	1
1	0

Boolean functions. Table 2.3 is a truth table that describes the behavior of two of the three basic operations: AND and OR.

The AND and OR operators are binary. Table 2.3, which is split into an input half and an output half, shows all possible values of the operands, a and b, and the results of applying the operator. From the table, we see that the AND operator is true, only if both of its operands are true. The OR operator is true if at least one of its operands is true.

Notice how we build a table with all possible combinations of the operands. In Table 2.3, we think of the operands a and b as forming a 2-bit number, with a as the high-order bit, and b as the low-order bit. We then count 2-bit numbers, starting at 0, and finishing when we can count no higher. Each count becomes a row in our table. The sequence counted in the above example is 00, 01, 10, 11, or in decimal 0, 1, 2, 3. These are the values on the rows entered in the input half. This counting technique can be applied to tables with any number k of inputs, by counting using k-bit numbers.

The NOT operator is unary. The truth table for the NOT operator is shown in Table 2.4. The truth table has only one input variable, and the NOT operator simply flips, or inverts, its value.

In addition to these three basic operators, Boolean algebra includes several non-basic, but common operators. We call these operators "non-basic" because they can actually be defined using the basic operators, so in a sense they are not atomic. These common operators are XOR, XNOR, NAND, and NOR.

TABLE 2.5   Truth table for XOR,
XNOR, NAND, and NOR.

a	b	$a \oplus b$	$a \odot b$	$\overline{a \cdot b}$	$\overline{a + b}$
0	0	0	1	1	1
0	1	1	0	1	0
1	0	1	0	1	0
1	1	0	1	0	0

They are all binary, and a truth table giving their definitions is found in Table 2.5. The XOR operator, indicated with the "$\oplus$" symbol is true if and only if its operands are different. The abbreviation XOR stands for *exclusive OR*. It also, equivalently, can be thought of as returning true if one of its operands is true, but not both. The XNOR function, indicated with the "$\odot$" symbol, returns true only if its operands are equal. We notice that XNOR is simply the XOR function inverted. NAND produces a true value only if one of its operands is false, or 0. It is the inverse of the AND function. The NOR function produces a true value only if its operands are all 0, and is the inverse of the OR function. In general, the "N" in the operator name implies an inversion of the named function.

2.2.2   Boolean Expressions and Truth Tables

As discussed previously, we form expressions using the operators from Section 2.2.1. The following equation gives an example.

$$g = (\overline{ab} + c) \oplus (a\overline{c} + \overline{b}) \tag{2.7}$$

You will notice, in Equation 2.7, that the AND operator is often omitted in expressions, and so when we see two operands adjacent to each other, with nothing between them, there is the implication of an AND operator. A similar notational shortcut is used with the multiplication operator in arithmetic algebra. Also, we assume that AND has precedence over OR.

Equation 2.7 uses a Boolean expression to define a function, $g(a, b, c)$, of three parameters. Looking at this equation, we understand how to compute the value of the function. But, we might be interested in seeing the results of applying the function g to any possible triplet of argument value. We then might consider trying to build a truth table for the function.

The technique used to build the truth table for complex functions involves building tables for larger and larger pieces of the defining expression, combining columns until the final column for the function is calculated. This is illustrated with Table 2.6, for the function in Equation 2.7. The table has three input variables, a, b, and c. Columns are built from left to right. Each

TABLE 2.6 Truth table for Equation 2.7.

a	b	c	ab	$\overline{ab}$	$\overline{ab}+c$	$\overline{c}$	$a\overline{c}$	$\overline{b}$	$a\overline{c}+\overline{b}$	g
0	0	0	0	1	1	1	0	1	1	0
0	0	1	0	1	1	0	0	1	1	0
0	1	0	0	1	1	1	0	0	0	1
0	1	1	0	1	1	0	0	0	0	1
1	0	0	0	1	1	1	1	1	1	0
1	0	1	0	1	1	0	0	1	1	0
1	1	0	1	0	0	1	1	0	1	1
1	1	1	1	0	1	0	0	0	0	1

TABLE 2.7 From table to equations.

a	b	c	h
0	0	0	0
0	0	1	0
0	1	0	1
0	1	1	0
1	0	0	0
1	0	1	1
1	1	0	1
1	1	1	1

column is built from previous columns, by applying a single operator. As we progress left to right we build larger and larger pieces of our formula. On the last column, we have both operands of the final XOR, and apply this operation to the columns for its operands, producing the column for g.

Using the above technique, we can build a truth table from any Boolean expression. It is also useful to be able to do the reverse. That is, it is useful to be able to construct Boolean equations from a truth table. We illustrate the technique used with an example. Consider the function defined by the truth table in Table 2.7. We wish to describe this function with a Boolean expression. To create the expression, we ask ourselves, what makes the function h true? To answer this question we might look down the output column for h, find all rows in the table where h is 1, and list them out in the following manner.

h is 1 only if a is 0, b is 1, and c is 0, or a is 1, b is 0, and c is 1, or a is 1, b is 1, and c is 0, or a is 1, b is 1, and c is 1.

We can see that the rows listed are the third, sixth, seventh, and eighth. If we write this statement using Boolean algebra, it would appear, much more compactly, as follows.

$$h = \overline{a}b\overline{c} + a\overline{b}c + ab\overline{c} + abc \tag{2.8}$$

TABLE 2.8 Don't
care truth table.

a	b	c	f	g
0	0	X	0	1
0	1	0	1	X
0	1	1	X	1
1	0	0	0	1
1	0	1	1	0
1	1	X	1	0

Expression 2.8 is composed of a series of what are called *minterms* being
ORed together. Each minterm comes from one row of the truth table, and
is the full set of independent variables ANDed together. The variables in the
AND operation are either unaltered values, if the original statement indicated
that they should be 1, or inverted values, if the original statement indicated
that they should be 0.

Minterms are often referred to by number. For example, we might refer to
Minterm 3. In the example from Table 2.7, Minterm 3 is the minterm with
variable values, that when interpreted as a binary number can be read as a
decimal 3. In this case the values of the variables would form the 3-bit binary
number 011, or written as a minterm expression, $\bar{a}bc$.

2.2.3 Don't Care Conditions

Another topic that needs to be covered in this section has to do with using
Boolean expressions to describe circuits. Often when describing a circuit, the
description is purposefully left incomplete. An analogy in arithmetic algebra
might be a description of the binomial coefficient function. A typical definition
of this function might be the following.

$$B(n, k) = \begin{cases} B(n-1, k) + B(n-1, k-1), & 0 < k \leq n \\ 1, & n = k \\ 1, & k = 0 \end{cases} \qquad (2.9)$$

An observer of our definition might read through the formulas, and when finished, observe the third case includes no mention of the independent variable
n, and so, it is incomplete. We might respond that if $k = 0$, the value of n
does not matter; the function B will return the value 1, regardless of the value
of n.

To allow incomplete specifications for circuits we introduce what are called
don't care values in truth tables. Consider the truth table in Table 2.8. In
this table, the symbol "X" is the don't care value. When read correctly, this

table gives us an incomplete description of the functions f and g. The first row reads as follows.

If a is 0, b is 0, and we don't care what c is, then the function f should output a 0, and the function g should output 1.

Notice that the don't care value allows this one row to specify the results of two rows in a full truth table. The variable c could either be 0 or 1, as a result of the don't care, and so this first row specifies the same function output for both minterm values 000 and 001. So, we see that don't cares, occurring on the input side of a truth table, allow for a shorter, more abbreviated table.

Reading the second row of Table 2.8, we might get the following specification.

If a is 0, b is 1, and c is 0, then the function f should output a 1, and we don't care what the output of g is.

When a don't care is found in the output of the function g, it specifies a case in which we do not expect to use the function with this specific set of inputs. When the function g is turned into circuitry, the circuit will, in fact, produce a value, if the inputs are 010. But, because we do not care what the output is in this case, as the designer, we can choose the output signal to our advantage. As designers, we are interested in producing as simple a circuit as possible, and so, the value we choose to output for a don't care is the value that produces the simplest circuit. We discuss this topic further, later in Section 2.2.5, when we discuss circuit simplification. But, what we have learned from examining the second row of the table, is that when a don't care value occurs in the output half of the table, that allows us to choose the output value to facilitate simplification.

2.2.4 Boolean Simplification Using Identities

In arithmetic algebra, we often perform formula simplification. We do this to reduce a formula to a form which is more readable, and from which we can more easily obtain significant information. As an example, consider the following expression.

$$(2a + 6) \cdot (2a - 6) \tag{2.10}$$

We might rewrite this as

$$(2a + 6) \cdot 2a - (2a + 6) \cdot 6 \tag{2.11}$$

We can make this change, because we are aware of what is called the *distributive law* of multiplication over addition. A second application of the distributive law would yield

$$2^2 a^2 + 6 \cdot 2a - (6 \cdot 2a + 6^2). \tag{2.12}$$

We could continue in this fashion, transforming our formula until we arrive at a form that satisfies us. For each transformation, we use a law, like the distributive law, to yield a new form. These laws are specified as identities. In other words, we use identities to replace pieces of the formula with other equivalent pieces. As an example of an identity, the distributive law is defined with the following identity.

$$a(b + c) = ab + ac \qquad (2.13)$$

We are also interested in simplifying Boolean expressions, in the same way we simplify arithmetic expressions, but not always for the same reasons. In processor design, Boolean expressions are transformed into digital circuitry. The more complex a Boolean expression, the more components the resulting circuit will contain. It is usually the case that the larger a circuit, the slower it will operate, the more electrical power it will consume, and the larger its physical size will be. So, our interest in Boolean expression simplification is not motivated by readability concerns, but rather concerns with how efficient the resulting circuit will be.

2.2.4.1 Boolean Identities

To simplify Boolean expressions, we need a set of identities, similar to the ones we use to manipulate arithmetic expressions. There a many such identities. However, below you will find some of the more useful ones.

1. Double negation: $\bar{\bar{a}} = a$

2. Contradiction: $a \cdot \bar{a} = 0$

3. Tautology: $a + \bar{a} = 1$

4. Commutativity: $a + b = b + a, \quad a \cdot b = b \cdot a$

5. Associativity: $a + (b + c) = (a + b) + c, \quad a \cdot (b \cdot c) = (a \cdot b) \cdot c$

6. Identity elements: $a + 0 = a, \quad a \cdot 1 = a$

7. Zero elements: $a + 1 = 1, \quad a \cdot 0 = 0$

8. Idempotency: $a + a = a, \quad a \cdot a = a$

9. Distributive: $a \cdot (b + c) = a \cdot b + a \cdot c, \quad a + bc = (a + b) \cdot (a + c)$

10. DeMorgan's: $\overline{a + b} = \bar{a} \cdot \bar{b}, \quad \overline{a \cdot b} = \bar{a} + \bar{b}$

11. Definition of XOR: $a \oplus b = \bar{a} \cdot b + a \cdot \bar{b}$

Many of these identities are familiar to anyone who has worked with arithmetic algebra. We see that if we think of OR as addition, and AND as multiplication, a lot of the rules in the two algebras are identical. There are some variations. For example, notice that several of the identities have two forms: one for AND, and one for OR. So, not only does AND distribute over OR, but OR also distributes over AND. There are also rules that are specific to Boolean algebra, such as idempotency, tautology, and contradiction.

2.2.4.2 DeMorgan's Law

DeMorgan's Law is a rule allowing us to do something very similar to what we do when we write

$$-(a + b) = -a - b \tag{2.14}$$

in arithmetic algebra. In this simplification we bring a negation into an operator group, and distribute it onto the terms of the group. DeMorgan's Law does the same in Boolean algebra. Notice that what happens when we distribute the negation, is that we invert the operands and flip the operator. That is to say, an AND becomes an OR, and an OR becomes an AND.

Another interpretation of DeMorgan's Law, is that it defines two forms of the NAND function, and two forms of the NOR function. These are useful definitions, on occasion, when we start building circuits from Boolean expressions.

2.2.4.3 Simplifying the XOR Function

Rule 11 is the definition of the XOR operator. This is the equation we get when we copy minterms from the XOR truth table. It is quite useful to remember this expanded form of the XOR function, when performing expression simplification. A corollary of this rule is that the XNOR function has an expanded form also.

$$a \odot b = ab + \bar{a}\bar{b} \tag{2.15}$$

Again, this is the formula we get when we copy minterms from the truth table for XNOR.

2.2.4.4 Example Simplification Using Identities

Let us now work through an example of how these identities might be used to simplify a Boolean function. Consider the following expression.

$$(a\bar{b} + c) \oplus \bar{b}c. \tag{2.16}$$

We can rewrite this expression as

$$\overline{a\bar{b} + c} \cdot \overline{b}c + (a\bar{b} + c) \cdot \overline{\overline{b}c} \tag{2.17}$$

using Rule 11. With two applications of Rule 10 we get

$$(\overline{a\overline{b} \cdot \overline{c}})(\overline{b} + \overline{c}) + (a\overline{b} + c)\overline{\overline{bc}}. \tag{2.18}$$

We then use Rule 1, to remove the double negation.

$$(\overline{a\overline{b} \cdot \overline{c}})(\overline{b} + \overline{c}) + (a\overline{b} + c)bc \tag{2.19}$$

A final application of Rule 10 yields the following.

$$(\overline{a} + \overline{\overline{b}}) \cdot \overline{c})(\overline{b} + \overline{c}) + (a\overline{b} + c)bc \tag{2.20}$$

An application of Rule 1 removes the double negation.

$$(\overline{a} + b) \cdot \overline{c}(\overline{b} + \overline{c}) + (a\overline{b} + c)bc \tag{2.21}$$

Using Rule 4 to rearrange terms results in

$$\overline{c} \cdot (\overline{a} + b)(\overline{b} + \overline{c}) + bc \cdot (a\overline{b} + c). \tag{2.22}$$

This is followed by two applications of Rule 9.

$$(\overline{c} \cdot \overline{a} + \overline{c}b)(\overline{b} + \overline{c}) + bca\overline{b} + bcc \tag{2.23}$$

In Expression 2.23 we see three terms being ORed together. We use Rule 8 to eliminate duplicate copies of c in the third term, Rule 4 to rearrange the second term, and Rule 10 to distribute in the first term.

$$(\overline{c} \cdot \overline{a} + \overline{c}b)\overline{b} + (\overline{c} \cdot \overline{a} + \overline{c}b)\overline{c} + ca b\overline{b} + bc \tag{2.24}$$

We now reverse some of the operands with Rule 4, use Rule 2 to determine that $b\overline{b} = 0$, and then use Rule 7 to eliminate the second-to-last term.

$$\overline{b}(\overline{c} \cdot \overline{a} + \overline{c}b) + \overline{c}(\overline{c} \cdot \overline{a} + \overline{c}b) + 0 + bc \tag{2.25}$$

Using Rule 6 and Rule 9, twice, we get

$$\overline{b}\overline{c} \cdot \overline{a} + \overline{b}\overline{c}b + \overline{c} \cdot \overline{c} \cdot \overline{a} + \overline{c} \cdot \overline{c}b + bc. \tag{2.26}$$

With applications of Rule 8 we eliminate duplicate variables, and using Rules 4, 2, and 7, we get

$$\overline{b}\overline{c} \cdot \overline{a} + 0 + \overline{c} \cdot \overline{a} + \overline{c}b + bc. \tag{2.27}$$

The zero term is eliminated with Rule 6, and we use Rules 4 and 9, twice.

$$\overline{c} \cdot \overline{a}(\overline{b} + 1) + b(\overline{c} + c) \tag{2.28}$$

From Expression 2.28, Rules 3 and 7 give us

$$\overline{c} \cdot \overline{a} \cdot 1 + b \cdot 1, \tag{2.29}$$

which, by two applications of Rule 6, yields a final result of

$$\overline{c} \cdot \overline{a} + b. \tag{2.30}$$

The rather lengthy derivation we have just completed is an exercise that shows us that algebraic simplification is difficult. It often involves many steps. It often requires the formulation of complex strategies. And, at every step, it often requires choices between alternative strategies, with little to help in the decision, other than the experience of the person performing the simplification.

It would be preferable if this simplification process could be automated. That is, it would be nice if this job could be done by a computer program. This program would take as input a Boolean expression, and output a simplified form of the expression. But the complexity of the process is going to make it quite difficult to write a program that does simplification algebraically. If we are, indeed, going to automate Boolean simplification, it is best to search for another method.

2.2.5 Boolean Simplification Using Karnaugh-Maps

Karnaugh-maps, also called K-maps, are a graphical way of representing a truth table. They make relationships between minterms stand out, allowing the minterms to be combined into groupings with simpler formulas.

The size of a K-map is determined by the number of independent variables in the function it represents. In our discussion we begin by considering the smallest number of independent variables first, and then work our way up to larger functions. The smallest number of variables a function can have is zero variables. There are only two functions of zero variable.

$$f_0 = 0 \tag{2.31}$$

$$f_1 = 1 \tag{2.32}$$

The expressions defining both of these functions are already as simple as possible, and so are not relevant to a discussion on simplification. Likewise for functions of one variable, there are also only two functions, and both of them have very simple definitions.

$$f_0 = \overline{x} \tag{2.33}$$

$$f_1 = x \tag{2.34}$$

So, the smallest interesting function is probably a function of two variables.

2.2.5.1 K-Map for Functions of Two Variables

Let's examine a function of two variables. An example is given by the truth table in Table 2.9. If we were to write a formula for the function g by copying the minterms that yield a value of 1, we would arrive at the following equation.

$$g = \overline{a}\overline{b} + a\overline{b} + ab \tag{2.35}$$

TABLE 2.9 Example
function of two variables.

a	b	g
0	0	1
0	1	0
1	0	1
1	1	1

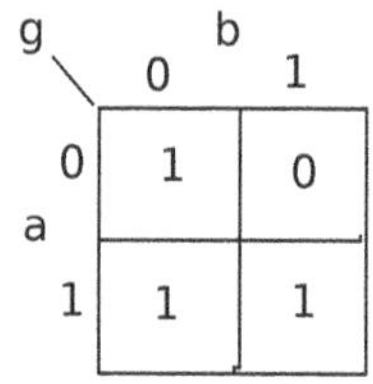

FIGURE 2.2 K-map for the function g of Table 2.9.

The K-map for this function would be drawn as shown in Figure 2.2. The K-map shows the minterms from the truth table as quadrants in a square. The square is a grid with two coordinates: the vertical coordinate is the first variable of the minterm, and the horizontal coordinate is the second variable of the minterm. Along the two axes are marked the variable names, and all possible values of the variables. To fill in the K-map, all entries belonging to minterms that make the function g true are filled in with a 1, and those making g false are filled in with a zero. For example, the minterm $a\bar{b}$ makes g a 1, so the cell with coordinates $a = 1$ and $b = 0$ is marked with a 1.

The K-map is filled in by copying in individual minterms. To simplify an expression, we copy out minterms in groups. Combining the minterms usually yields simpler expressions. The rules for combining minterms into groups can be stated as follows.

1. Adjacent cells that contain 1 can be combined.

2. Combined cells must form a rectangular group.

3. The size of a group must be a power of two.

4. The groups copied out must cover all cells that are 1. (Notice, however, that cells may be covered by several groups.)

FIGURE 2.3 Grouping for the function g from Table 2.9.

TABLE 2.10   Example three-variable function.

a	b	c	h
0	0	0	1
0	0	1	0
0	1	0	1
0	1	1	0
1	0	0	1
1	0	1	1
1	1	0	1
1	1	1	0

When copying out groups, the larger the groups, the more simplification will occur. Also, the fewest groups result in a simpler expression. In the example of Figure 2.2, the largest group satisfying our criteria is of size two. In fact, all of the cells with 1 in them can be covered by two groups of two cells. The two groups are shown in Figure 2.3. To use these groupings to simplify g, we come up with Boolean expressions describing the two groups, and OR the two expressions together. In Figure 2.3, we see that in the vertical group, the value of a changes. To describe the group we would turn to the variable b, which we see is 0, throughout the group. That b is 0, would be written as $\bar{b}$. The horizontal group is analyzed in a similar way. Notice that, in this group b changes, and so it is not useful to describe the group. The horizontal group would then be described by saying that a is a 1, or as the Boolean expression, a. Combining the descriptions of the two groups, we get the simplified formula

$$g = a + \bar{b} \tag{2.36}$$

as the equation for the two groups. This is the simplified formula.

2.2.5.2   K-Maps for Functions of Three Variables

Let us now turn to functions of three variables. One such function is defined by the truth table in Table 2.10. Three-variable K-maps are formed by placing

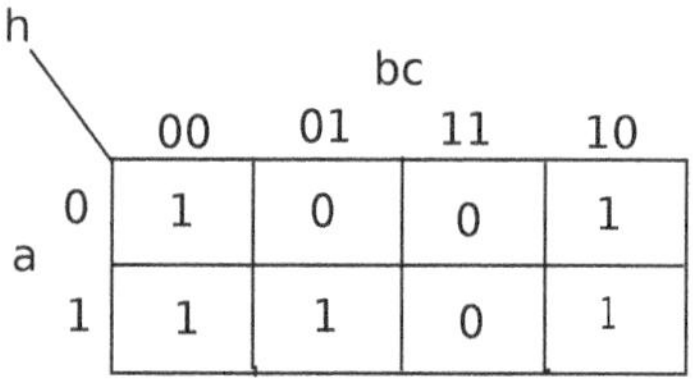

h

	bc			
	00	01	11	10
a 0	1	0	0	1
1	1	1	0	1

FIGURE 2.4 K-map for the function h from Table 2.10.

two 2-variable K-maps side by side. For Table 2.10, this would give us the
K-map shown in Figure 2.4. This map has, again, the first variable of the
minterm on the vertical axis. There are two variables remaining, and so they
are installed as a 2-bit coordinate along the horizontal axis. Notice the order
of coordinate values on the horizontal axis. In decimal, the order would be
0, 1, 3, 2. This order is important. It is called the *Gray code* sequence, and
has the property that, between consecutive counts in the sequence, only one
bit changes. For instance, between 0 (00) and 1 (01), only the low-order bit
changes. Between 1 (01) and 3 (11), only the high-order bit changes. The same
is true between 3 (11) and 2 (10), and also between 2 (10) and 0 (00). This
calls to attention that we consider the cell with coordinate 10 to be adjacent to
the cell with coordinate 00. Karnaugh-maps wrap around, both horizontally
and vertically.

FIGURE 2.5 Grouping for the function h from Table 2.10.

Using the 3-variable K-map to simplify h, uses the same procedure as we
used in the two-variable K-map. The largest group found in this example is
the group of four, split between the first and last column. There are no further
groups of four, and so the last two cells with values of 1 are covered by a group
of size 2. The grouping is shown in Figure 2.5. The group of four is described
by the expression $\bar{c}$ (c is the only unchanged variable in the group), and the
group of two is described by $a\bar{b}$. This yields the following equation for h.

$$h = c + a\bar{b} \tag{2.37}$$

TABLE 2.11 Example function of four variables.

a	b	c	d	z
0	0	0	0	1
0	0	0	1	1
0	0	1	0	1
0	0	1	1	1
0	1	0	0	1
0	1	0	1	0
0	1	1	0	1
0	1	1	1	0
1	0	0	0	1
1	0	0	1	1
1	0	1	0	1
1	0	1	1	1
1	1	0	0	0
1	1	0	1	0
1	1	1	0	1
1	1	1	1	0

z		cd		
ab	00	01	11	10
00	1	1	1	1
01	1	0	0	1
11	0	0	0	1
10	1	1	1	1

FIGURE 2.6 K-map for the function z from Table 2.11.

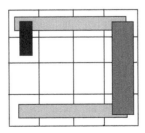

FIGURE 2.7 Grouping for the function z from Table 2.11.

TABLE 2.12 K-map with don't care conditions.

a	b	c	m
0	0	0	1
0	0	1	X
0	1	0	0
0	1	1	1
1	0	0	X
1	0	1	1
1	1	0	X
1	1	1	0

2.2.5.3 K-Maps for Functions of Four Variables

Karnaugh-maps for four variables are formed by stacking two three-variable K-maps. Consider the function of four variables with its truth table given in Table 2.11. Its K-map would be as shown in Figure 2.6. A group of eight is split between the first and last rows. A group of four covers the rightmost column, and the last cell containing a 1 can be covered by a group of two. This is shown in Figure 2.7. Copying out the three groups, in the usual way, yields the following,

$$z = \bar{b} + c\bar{d} + \bar{a} \cdot \bar{c}\bar{d} \tag{2.38}$$

where $\bar{b}$ describes the group of eight, $c\bar{d}$ describes the group of four, and $\bar{a} \cdot \bar{c}\bar{d}$ describes the group of two.

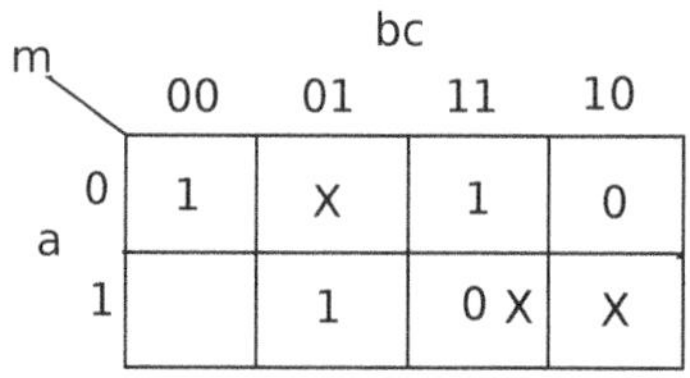

m \ bc	00	01	11	10
a 0	1	X	1	0
1		1	0 X	X

FIGURE 2.8 K-map for the function m from Table 2.12.

2.2.5.4 Don't Care Conditions in Karnaugh-Maps

Returning to the topic of don't cares, these values can be tremendously valuable when simplifying with Karnaugh-maps if they occur for output values. This is illustrated by the example with the truth table given by Table 2.12. When we build the K-map for this function it will contain cells with 0 values, 1 values, and don't care values. This K-map is shown in Figure 2.8. Without

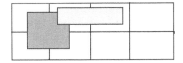

FIGURE 2.9 Grouping for the function m from Table 2.12.

using the don't care values, we would be required to pull out each cell as a group of size one, producing the formula

$$m = \bar{a}\bar{b}\bar{c} + \bar{a}bc + a\bar{b}c \tag{2.39}$$

which is not at all simplified. But, remember that we, as designers, can pick don't care values to help in simplification. So, we can choose values of 1 for Minterm 1 and Minterm 4, and choose a value of 0 for Minterm 6. This results in coverage with a group of four and a group of two, shown in Figure 2.9. This grouping results in the much simpler equation

$$m = \bar{b} + \bar{a}c. \tag{2.40}$$

2.2.5.5 K-Maps for Functions of More than Four Variables

We could, now move up to consider K-maps for functions of five or more variables. These maps could be drawn and analyzed. However, the K-maps rapidly get large, and producing them by hand becomes burdensome.

Remember why we developed K-maps in the first place. We did this so that we would have available a method that was much more mechanical than simplification using algebra. K-maps are, in fact, mechanical enough in their use so that we can write computer programs to produce and use them. And, when we start climbing up to Karnaugh-maps beyond four variables, it is more time efficient to use one of the K-map programs, rather than hand produce the K-map.

2.3 SUMMARY

This chapter presents the basic mathematical material required to design a processor. The major topics that are requisite are those of the binary number system and Boolean algebra.

Binary is ubiquitous in computer design. As the choice of representation for numeric data in the modern computer, it is difficult to imagine discussing computer organization and architecture without knowledge of this system. In binary, we use a base of two for representing numbers, as opposed to a base of ten, which is used in the much more familiar decimal system. Using two's compliment, in conjunction with binary, allows us to represent both non-negative and negative numbers.

Boolean algebra is used extensively in circuit design. It is an algebra based on two states, often referred to as 0 and 1. It is the model upon which computer circuitry is built.

There is a connection between Boolean algebra and binary. They are both constructed around values of 0 and 1. As a matter of fact, it is not to far-fetched to think of binary numbers as simply a multi-bit version of a Boolean value.

2.4 EXERCISES

2.1 Write the following integers in (1) decimal, (2) in 8-bit binary, and (3) in hexadecimal.

   a. 109 decimal.

   b. 00010001 binary.

   c. A5 hexadecimal.

   d. 95 decimal.

   e. 76 hexadecimal.

   f. 11100011 binary.

2.2 Do the following calculations, using binary addition.

   a. 10110110 + 11101010

   b. 00011100 + 00101101

   c. 01110111 + 01001100

2.3 Give the 12-bit (1) sign-magnitude representation, (2) one's compliment representation, and (3) two's compliment representation of the following signed integers.

   a. 149

   b. −149

TABLE 2.13   Truth table for
Exercise 2.8.

i	m	n	g
0	0	X	X
0	1	0	1
0	1	1	0
1	X	0	X
1	X	1	1

c. −208

2.4 Draw the truth tables for the following Boolean functions.

   a. $f = \overline{a + b} \cdot (cb + \bar{a})$

   b. $f = (a \oplus b) + (c \oplus b)$

   c. $f = (abc + \bar{a}\bar{b}c)(\bar{a}\bar{b} + \bar{c})$

2.5 Simplify the following Boolean expressions using the identities given to you in Section 2.2.4.1.

   a. $\overline{\overline{a + b} \cdot (c \oplus b)}$

   b. $\overline{(a + b)(\bar{a} + \bar{b})} \oplus (a \oplus b)$

   c. $(\bar{c} + (\bar{b} + \bar{a})(b + c))a$

2.6 Use K-maps to simplify the following Boolean functions.

   a. $f = a\bar{b}\bar{c} + \bar{a}bc + b\bar{c} + \bar{b}\bar{c} + a\bar{b}c$

   b. $f = ac + \bar{a}\bar{b}\bar{c} + \bar{a}\bar{b}c + ab\bar{c}$

   c. $f = \bar{a}\bar{b}\bar{c} + abc + a\bar{b}c + \bar{a}b\bar{c}$

2.7 Use K-maps to simplify the following Boolean functions.

   a. $g = \bar{x}y\bar{z}w + xy\bar{z}w + x\bar{z} \cdot \bar{w} + \bar{x} \cdot \bar{z} \cdot \bar{w} + \bar{x} \cdot \bar{y}zw + x\bar{z}w$

   b. $g = xyz + \bar{x} \cdot \bar{y}z + \bar{x}y\bar{z} \cdot \bar{w} + x\bar{y}z + \bar{x}y\bar{z}w$

2.8 Given the truth table of Table 2.13, give the simplified equation for g, using K-maps.

Digital Circuitry

CONTENTS

The question we are examining is, "How do computers do what they do?" This question is difficult to answer at this point. But, one way to tackle this question is to start by describing the pieces, or components, of a computer, and how they interact.

We tackle the question of how processors do their work, starting at the very bottom level. We then work our way up, level by level, until we understand the full workings of a modern processor. We start at the digital circuitry level.

Processors are *digital circuits*. A digital circuit is an electrical circuit in which voltage is treated as a signal. These signals carry information, and the components of the circuit process the information. The number of signals is limited. Each individual signal is a distinct voltage, and so digital circuits carry only a finite set of discrete voltages. When we limit the number of signals, we ensure that processing will be relatively easy, and yet we need a sufficient number of different signal values, so that all information that the processor works with can be represented.

So, now we have come to another question: "What information do processors work with?" The answer is processors work only with numbers. This may be a little hard to believe, since you have, no doubt, worked with non-numeric material on a computer, such as with a word processor, working with natural language, or a audio player working with acoustic signals. But the fact is that, yes, computers only work with numbers. Later, in Section 6.1.2.1, we explore how this is possible.

The second question, motivated by the above discussion, is how many signal values do we need to represent all of the material the processor works with? The answer to this question is, two signals. All numbers can be represented with two signals.

Signals on a computer are represented as voltages. A common scheme is to use a two-voltage system, with 5V used as one signal, and 0V used as a second signal. These two signals might simply be referred to as the *high* and *low* signal, respectively. From our perspective, the actual voltage value of the signal is irrelevant. Also, that one signal is a high voltage and the other is a low voltage is irrelevant. These two factors are irrelevant because the voltage value is not what carries information. Rather, what conveys information is that a wire is carrying one of two distinct signals. It is therefore common just to refer to the two signals as the value 1 or as the value 0. Note that it is immaterial if the high signal is interpreted as a 1, and the low signal as 0, or the low signal is interpreted as a 1, and the high signal as 0. In fact, circuits are built following both of these conventions.

The two signals, 0 and 1, can be used to represent all numbers using the binary number system.. But, there is another interpretation of the two digital signals. We often view them as representing the Boolean values of *True* and *False*, with 0 used as *False*, and 1 as *True*. It is this Boolean view that is most useful in circuit design.

A digital circuit can usually be classified in one of two categories.

- Combinational circuitry.

- Sequential circuitry.

It is difficult, at this point, to describe the difference between the two types of circuitry. A good description requires more background work. However, an intuitive explanation might emphasize that a sequential circuit has memory, and a combinational circuit does not. The result of this is that the outputs of a combinational circuit change on every input. This may not be true for a sequential circuit, that remembers its previous value, and clings to this old value, and so does not change its output.

This chapter discusses both combinational and sequential circuitry. We begin with a discussion of combinational circuitry. We begin here because combinational circuitry is, generally, conceptually easier to understand than sequential circuitry. The second part of the chapter covers sequential circuitry.

3.1 COMBINATIONAL CIRCUITS

After our introduction to Boolean algebra in Chapter 2, we are now ready to start using it to build combinational circuits.

3.1.1 Designing with Logical Gates

A combinational circuit is a direct hardware implementation of a Boolean function. Combinational circuits consist of components that compute operations, and connections that carry Boolean signals to and from the operator components.

The components of a digital circuit are called *gates*. Gates are built to do all of our common Boolean operations. The symbols for the gates that we are

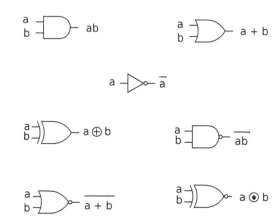

FIGURE 3.1 Different gate types.

interested in are shown in Figure 3.1. They compute the indicated functions: AND, OR, NOT, XOR, NAND, NOR, and XNOR. Each gate has input pins, located on the left, and an output pin located on the right. The gates are used by connecting signals of either 0 or 1 on to the input pins. The output pin then shows the result of the given function for those inputs. For example, if the inputs for the AND gate were $a = 0$ and $b = 1$, the gate would produce the result of the AND of these two values, 0.

The common gates can be wired together to compute arbitrary Boolean functions. Consider the following example function.

$$f = (a \oplus b)(\bar{b} + c) \tag{3.1}$$

Each operation in the expression is implemented by one of the common gates.

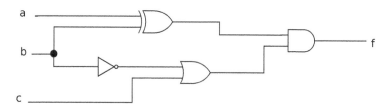

FIGURE 3.2 Schematic of function f from Example 3.1.

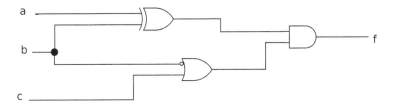

FIGURE 3.3 Schematic of f from Example 3.1, with abbreviated invert-ers.

The gates are wired together to feed an output of one operation as the input to another appropriate operation. So, this function would be wired as in Figure 3.2. A more common way of drawing this circuit would be what is shown in Figure 3.3. Here, the NOT gate, better known as an *inverter*, has been replaced by a small open circle on the base of the OR gate. This small circle indicates inversion, as can also be seen on the NAND and NOR gates, where it is used on the output.

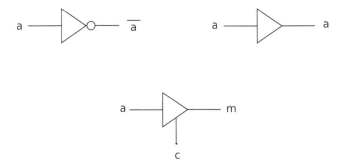

FIGURE 3.4 Different buffer types.

We might wonder, if the small circle on the inverter indicates inversion, then what is the purpose of the huge triangle? This question brings up another set of useful digital components, called *buffers*. The triangle on the inverter is a buffer. There are three buffer types that are in common use in digital circuitry. The symbols for the three types are shown in Figure 3.4. We are already familiar with one of these buffers: the *inverter*. As shown, it takes an input, and outputs the inverse of the input. The second buffer, shown on the right, is a *simple buffer*. It has no inversion element, and as indicated, outputs its input without modifying its truth value. It might be a little baffling trying to work out what this device is used for. The answer to this question has to do with *signal fan-out*.

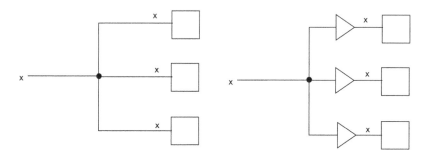

FIGURE 3.5 Fan-out illustration.

In digital circuits, it is quite common to split a single signal, as shown in Figure 3.5, on the left. This splitting of a signal is called fan-out. The problem with fan-out is that the signal on each branch of the split is weakened. The strength of the original signal is divided among the branches. With a lot of fan-out, the original signal may become so weak as to be unusable.

Although the simple buffer does nothing, logically, to a signal, it does, however, boost the signal back up to full strength. Figure 3.5, to the right, shows how buffers might be used to restore a signal after fan-out.

The last type of buffer shown in Figure 3.4 is known as a *tri-state switch*. As the name implies, it is a switch. When it is closed, the input signal, a, will show on the output pin, m. When open, the output m shows no signal at all. The pin m is put into what is known as a state of high impedance. In layman's terms, the power to m has been turned off.

The tri-state switch is controlled by the control input c. When $c = 0$, the switch is open, and the signal a is stopped. When $c = 1$, the switch is closed, and the signal output at m will be the value of the input a. The name of the tri-state switch indicates that it is not really a Boolean component. This is because the output can actually present one of three possible values, and not one of two values, as does a true Boolean component. The switch can either be 0, 1, or Z. The symbol Z is often used to denote the state of high

TABLE 3.1   Tri-state switch.

a	c	m
0	0	Z
0	1	0
1	0	Z
1	1	1

impedance. The behavior of the tri-state switch is summarized by the tri-state table shown in Table 3.1.

3.1.2   Common Combinational Circuits

We are now ready to begin designing the low-level combinational components of the processor. The first component we will design is called a *decoder*.

3.1.2.1   The Decoder

An analogy to the operation of a decoder might be a hypothetical room, with a keypad on the wall. The keypad accepts numbers between zero and three. In the room are four lights: a floor lamp, a room light, a plant light, and a desk light. If the number 0 is entered at the keypad, the floor lamp lights up. All of the other lights are turned off. If a 1 is entered, the room light comes on, and all other lights are extinguished. If the number 2 or 3 is entered a similar action occurs with the plant light and the desk light, respectively. What we are describing is a four-way switch, operated by the entry of a code.

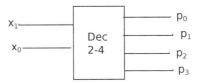

FIGURE 3.6   Decoder interface.

The decoder has an interface, that is illustrated by the *block diagram*, or *pin-out diagram*, in Figure 3.6. These diagrams show the external connections to a circuit, and in this way show the interface to the component.

The inputs x_1 and x_0, from Figure 3.6, form a 2-bit binary number, with x_1 as the high-order bit, and x_0 as the low-order bit. This number can be

TABLE 3.2 Decoder truth table.

x_1	x_0	p_0	p_1	p_2	p_3
0	0	1	0	0	0
0	1	0	1	0	0
1	0	0	0	1	0
1	1	0	0	0	1

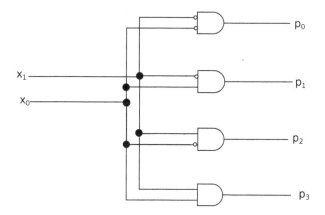

FIGURE 3.7 Schematic of a decoder.

between 0 and 3, corresponding to one of the four indexes on the p outputs. If the code on the x inputs is j, then output p_j will present a 1 signal, and the other p outputs will present 0 signals. This behavior is described in the truth table for the decoder in Table 3.2. We follow our normal design method for combinational circuits to build the decode, namely we copy out equations from the truth table, simplify them if necessary, and implement them using gates. We first copy out the equations. We need one equation per output. Examining the table, each p output column contains only one minterm, so the equations derived from the truth table are already quite simple.

$$p_0 = \overline{x_1} \cdot \overline{x_0} \qquad (3.2)$$

$$p_1 = \overline{x_1} x_0 \qquad (3.3)$$

$$p_2 = x_1 \overline{x_0} \qquad (3.4)$$

$$p_3 = x_1 x_0 \qquad (3.5)$$

Wiring this up gives us the circuit diagram of Figure 3.7.

TABLE 3.3 Encoder truth table.

p_0	p_1	p_2	p_3	x_1	x_0
1	0	0	0	0	0
0	1	0	0	0	1
0	0	1	0	1	0
0	0	0	1	1	1

The decoder of Figure 3.6 has a size indicated: 2-4, giving the number of inputs as the first value, and the number of outputs as the second value. Decoders come in different sizes, although the sizes all have the same form: $k - 2^k$. k is the number of input lines. With k input lines, an index value between 0 and $2^k - 1$ can be specified, and so the number of outputs is be 2^k.

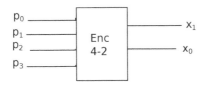

FIGURE 3.8 Encoder interface.

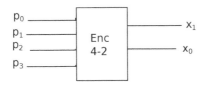

FIGURE 3.9 K-maps for an encoder.

3.1.2.2 The Encoder

A component that does the opposite of the decoder is the *encoder*. It can be thought of as a device that checks several circuits, and reports a code, indicating which of them is current active. The interface for a 4-2 encoder is shown in Figure 3.8. To use the encoder, one, and only one, of the input lines p is set to 1. The others are held at 0. The encoder then produces the index of the line that has the 1 signal. That is, if $p_j = 1$, and all other inputs are 0, the

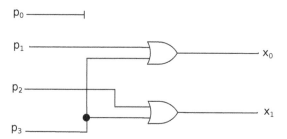

FIGURE 3.10 Encoder circuit.

output would be j, in binary. This behavior is summarized in the truth table given in Table 3.3. This table is an abbreviated table. Notice that because there are four inputs to the encoder, a full table should have sixteen rows. Table 3.3 has only four, showing only the inputs considered legitimate. The other rows of the full table would have don't care outputs, a fact we will use in our K-maps. These Karnaugh-maps are shown in Figure 3.9. An examination of these Karnaugh-maps reveals that the map for x_1 can be covered using a group consisting of the two rightmost columns, and the two center columns. Covering the K-map for x_0 is best done by using the two middle rows, and the two middle columns. The resulting formulas are the following.

$$x_1 = p_2 + p_3 \tag{3.6}$$

$$x_0 = p_1 + p_3 \tag{3.7}$$

These equations are implemented with the circuit shown in Figure 3.10. Notice that p_0 is not used, and is left unconnected.

The encoder we have designed is a very simple encoder, and requires the user to use it correctly. If the encoder is given several input signals that are 1, it will produce a code. However the code is meaningless. So, the user must be vigilant when wiring up the encoder. Some encoders do accept several input lines with 1 values. These encoders, called *priority encoders*, typically output the code for the 1-asserted input line with the highest index.

As with decoders, encoders come in various sizes. The size of the encoder of Figure 3.8 is a 4-2, specifying two inputs, and four outputs. In general an encoder has a size of 2^j-j.

TABLE 3.4   MUX truth table.

i_0	i_1	i_2	i_3	s_1	s_0	p
0	X	X	X	0	0	0
1	X	X	X	0	0	1
X	0	X	X	0	1	0
X	1	X	X	0	1	1
X	X	0	X	1	0	0
X	X	1	X	1	0	1
X	X	X	0	1	1	0
X	X	X	1	1	1	1

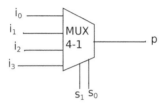

FIGURE 3.11   MUX interface.

3.1.2.3   The Multiplexer

The next component we examine is the *multiplexer*, or, as it is often referred to, the *MUX*. A block diagram of a 4-1 MUX is shown in Figure 3.11. A MUX can be thought of as a router. With the 4-1 MUX, there are four i inputs, trying to get through to the output p. Only one is allowed through, and the others are stopped. The selection of the output is done using the s selector lines. Collectively the s lines form a 2-bit binary number. This number specifies the index of the input i line to be let through to the output. We have summarized the behavior of the MUX in Table 3.4.

To arrive at an equation for the output p, we could plot the p on a K-map, and copy out the equations. A much shorter, but equally valid method is to build the equation by copying out rows from the truth table, as if they were minterms, ignoring the don't care conditions. So, for example, the pseudo-minterm specified by the second row in the table would be $i_0 \overline{s_1} \cdot \overline{s_0}$. Notice that the input variables with don't care conditions have been omitted from the pseudo-minterm name, since they are not being used to determine the value of the output p. Continuing in this fashion, we observe that p is true only on four rows of Table 3.4, yielding the equation

$$p = i_0 \overline{s_1} \cdot \overline{s_0} + i_1 \overline{s_1} s_0 + i_2 s_1 \overline{s_0} + i_3 s_1 s_0. \tag{3.8}$$

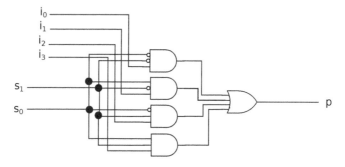

FIGURE 3.12 Schematic for a MUX.

The schematic for the circuit implementing Equation 3.8 is found in Figure 3.12.

Like decoders and encoders, multiplexers come in various sizes. The MUX shown in Figure 3.11 is a 4-1 MUX, indicating that there are 4 inputs and one output line. For now, all multiplexers will have only one output, so the second part of the size specification will always be 1. In general, MUX sizes will be of the form $2^j - 1$, with j selector lines.

3.1.2.4   MUX Composition

It is not uncommon to require a large MUX, but only have at your disposal a collection of smaller multiplexers. There is a fairly systematic way of combining multiplexers to increase the size. This technique is best illustrated with an example. Suppose that we need a 4-1 MUX, but possess only 2-1 multiplexers. We might combine three 2-1 multiplexers to create the 4-1, using the structure shown in Figure 3.13.

It is instructive to think of this structure as a tournament between four players. The players compete in pairs, in the first round, and the winner moves on to the final. s_0 is used to decide between the odd-numbered and even-numbered player, in the first round. In the final heat, s_1 selects which of the two preliminary heats contains the winner. A value of 0 selects the winner from i_0 and i_1, and a value of 1 selects the winner from i_2 and i_3. Each input wire i_k has been wired into the preliminary multiplexers based on weather k is odd or even, and weather the upper bit of k is 0 or 1.

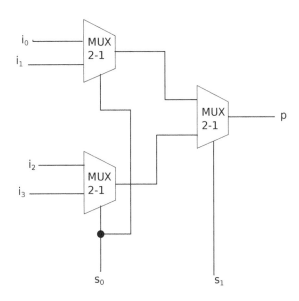

FIGURE 3.13 Interleaving MUXs.

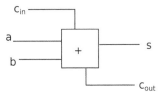

FIGURE 3.14 Adder interface.

TABLE 3.5 Adder truth table.

c_{in}	a	b	c_{out}	s
0	0	0	0	0
0	0	1	0	1
0	1	0	0	1
0	1	1	1	0
1	0	0	0	1
1	0	1	1	0
1	1	0	1	0
1	1	1	1	1

3.1.2.5 The Adder

We now move on to the next component: the *adder*. An adder is used to add together 1-bit numbers. It performs the following type of problem.

$$
\begin{array}{rcc}
c_{in} & 0 & 1 \\
a & 0 & 1 \\
+b & +1 & +1 \\
\hline
sc_{out} & 01 & 11
\end{array}
\tag{3.9}
$$

Here we show the addition of three bits: a carry-in, c_{in}, a, and b, producing a 2-bit answer, consisting of a sum bit, s, and a carry-out bit, c_{out}. A couple of examples are also shown. The interface to this circuit would appear as in Figure 3.14.

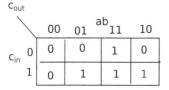
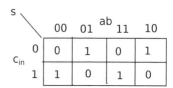

FIGURE 3.15 K-maps for the adder outputs.

We design this circuit using the usual design method for combinational circuits. First we fill in the truth table of the circuit. This is done simply by calculating the results of the addition problem specified by each row of the table. The resulting truth table is shown in Table 3.5. From the table we build the K-maps for c_{out} and s, shown in Figure 3.15. From the K-maps we copy out the equations for c_{out} and s.

$$
s = c_{in} \oplus a \oplus b
\tag{3.10}
$$

$$c_{out} = c_{in}a + c_{in}b + ab \tag{3.11}$$

The equation for c_{out} is derived by using the K-map in the usual way. The equation for s, however, requires a bit of explanation.

Let's go back and re-examine the XOR and XNOR functions. When we presented these functions, we explained that the XOR function outputs true if and only if its two operands are not equal. The XNOR function produces true only if its two operands are equal. Here, however, we are faced with the XOR of more that two operands. We might wonder, how does the XOR, or for that matter the XNOR function, extend to more than two operands? The answer to this question is that the XOR becomes the function commonly called the *odd function*, and the XNOR becomes the *even function*. With more than two operands, the XOR function, essentially, counts the number of its operands that have a value of 1. It outputs a true value if and only if an odd number of operands have a value of 1. Equivalently, we might say that XOR outputs 1 only if its operands have *odd parity*. In a similar way XNOR outputs 1 only if an even number of its operands are 1, or equivalently, its operands have *even parity*.

The Karnaugh-maps for the XOR and XNOR functions have a distinctive appearance. Both functions produce a checker-board pattern, as displayed in the K-map for s of Figure 3.15. The two checker-board patterns are duals of each other, meaning that the cells that are 0 for XOR, are 1 for XNOR, and vice versa. The minterms with 1 in the cell, for the K-map for s, have coordinates 100, 001, 111, and 010. All of these have odd parity, and so we know that the function s is simply an XOR of the three input variables.

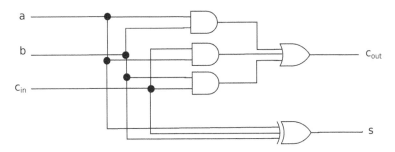

FIGURE 3.16 Schematic for the adder.

The circuit for the adder, built from Equations 3.10 and 3.11, is shown in Figure 3.16.

3.1.2.6 The Ripple-Carry Adder

The problem with our adder circuit is that it only performs arithmetic on very small numbers: 1-bit numbers. We, most likely, would like to perform arithmetic on larger, multi-bit numbers. For example, we might be interested in performing the following 4-bit addition.

$$
\begin{array}{r}
{\scriptstyle 1\,1\,1\,1\,0} \\
1011 \\
+1101 \\
\hline
1000
\end{array}
\qquad (3.12)
$$

Fortunately we have built an adder circuit that can be combined easily into a larger circuit that can add 4-bit numbers.

The circuit for 4-bit addition that we are building is called a *ripple-carry adder*. Ripple-carry adders can be built to any size. We will stick with Example 3.12, and build a 4-bit ripple-carry adder.

To build the ripple carry adder, we use our 1-bit adders, one adder to add each column in Computation 3.12. In each column, j, a carry-in, $c_{in,j}$ is added to an operand a_j and an operand b_j. The result is a sum bit s_j, and a carry-out, $c_{out,j}$. The carry-out, $c_{out,j}$, becomes the carry-in for the next column, $c_{in,j+1}$. It is by design that our 1-bit adder adds a carry-in, an a operand, and a b operand, producing an s

Following the guidelines just laid out, we might produce a 4-bit ripple-carry adder, as shown in Figure 3.17. In Figure 3.17, we have labeled the adders with subscripts, showing the columns they add, starting with the rightmost column, Column 0, and ending with the leftmost column, Column 3. The block diagram on the right shows the interface for our ripple-carry adder. In the interface diagram, several pins have slashes drawn through them, with numbers next to the slashes. These ports are called *buses*. They represent multi-bit inputs or outputs. The number on the slash gives the width of the bus. For example, one bus in the diagram is labeled as a, and has width 4. This indicates that a set of 4 input wires enter, labeled a_0–a_3, as indicated in the diagram on the left.

The diagram on the left of Figure 3.17 illustrates well why this circuit is called a *ripple-carry* adder. Each adder computes its results, but requires the carry-out from the next lower bit. In fact, quite accurately, we could say that an adder has to wait for the next lower adder to complete its result, before it can correctly compute its own result. In this way we can view the computation as rippling up the ripple-carry adder, following the propagation of the carry-out bits, from the low-order bit up to the high-order bit.

sum bit, and a carry-out, just as required.

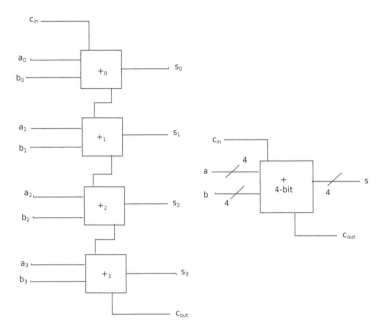

FIGURE 3.17 A ripple-carry adder.

3.2 SEQUENTIAL CIRCUITS

Remember that a sequential circuit includes memory, allowing the circuit to remember, or store, a value from the past. The value stored in a sequential circuit is called the *state* of the circuit. Sequential circuits are called *sequential*, because we think of them as moving through a sequence of states. To understand this statement, consider an analogy, using an example from computer programming, and illustrated by the following code sequence.

```
sum = 0
for i = 1 to n do
      sum = sum + i
```

The code uses two variables, i and sum. These variables store values, and these values form the state of the computation. The initial state is $(i = 1, sum = 0)$. After the first loop iteration the state becomes $(i = 2, sum = 1)$. After the second iteration the state is $(i = 3, sum = 3)$, and so on. As with a sequential circuit, we can think of the computation as moving from one state to another, over time. Notice that we now have the concept of time. This is in contrast with combinational circuits that compute an instantaneous result.

3.2.1 The Clock

When the concept of state is introduced into circuitry, as mentioned above, the concept of time becomes relevant. A device that delivers time to a circuit is called a *clock*. Other than keeping track of time, clocks are very different from the wall clocks we are used to.

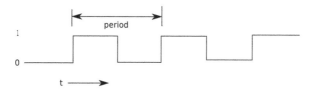

FIGURE 3.18 Timing diagram showing the clock signal.

A clock is a device that produces a signal, best explained using the timing diagram in Figure 3.18. The diagram shows the output of the clock as it varies over time. The shape of the output signal is a square sine wave. The wave is regular, with the time in which the clock produces a 0 signal equal to the time in which it produces a 1 value. Thinking of the clock signal as a sine wave, it makes sense to talk about the *cycle* of the clock, which consists of a *rising edge*, at which time the clock signal goes from 0 to 1, a peak area where the signal is 1, a *falling edge*, where the signal drops down to 0, and a trough area, where the signal is a 0. The amount of time that passes from the beginning of one cycle to the beginning of another cycle is called the *period*. The inverse of the period is the *frequency* of the clock, defined by the formula $F = \frac{1}{P}$, where P is the period, and F is the frequency.

The unit of measure of frequency is cycles/seconds. This unit is also, more compactly, called a *Hertz*, and abbreviated *Htz*. A clock that operates at 50 MHtz, then, is a clock producing 50,000,000 cycles per second.

We now, momentarily, turn back to the design of a processor, in order to explain how the clock is used. A processor is a collection of connected digital circuits; some of them are combinational, and some are sequential. The sequential circuits change state over time. If these circuits are allowed to change state at will, synchronization problems arise. The job of the clock is to synchronize the sequential circuits in the processor. What this means is that when we design a processor, we try to ensure that all sequential circuits, that have been selected for state change, change state simultaneously, and only at specific points of time. These points of time are defined by the clock cycle.

A clock cycle has two edges, as mentioned: the rising edge, and the falling edge. When designing a processor, we pick one of the two edges as the *trigger edge*, also called the *clock pulse*. This is the point of time at which all sequential circuits in the processor will change state. It is quite typical to use the rising edge, but there are plenty of circuits built with the falling edge as the trigger edge. To simplify our discussion, in this text we will assume that the rising edge is the trigger edge.

Our view of processor synchronization is now fairly simple; all circuits change state at the trigger edge of the clock cycle. Notice that in our simplified world, any state change must wait until the next trigger edge shows up. However, there are situations in which we want some piece of circuitry to change state immediately. Because of this, we find a small number of circuits that change state at will. They are not driven by the clock. We use the terms *clocked circuits* and *unclocked circuits* to refer to the two types of circuitry, where clocked circuits are synchronized to the clock and unclocked circuits are not.

3.2.2 Storage Devices

In order to build a sequential circuit, we need a device that can store a value. To keep things simple, we might like a device capable of storing a single bit. If so, we have several options. Firstly, we have devices that are unclocked, called *latches*, and devices that are clocked, called *flip-flops*. And secondly, there are multiple types of latches and flip-flops, suited to different applications.

3.2.2.1 The D-Type Storage Devices

We start our description of storage devices with the D-type device. We have both D-latches and D-flip-flops available, so we will begin with the simpler device, the D-latch.

FIGURE 3.19 D-latch interface.

TABLE 3.6 D-latch excitation table.

D	C	$Q_{(1)}$
X	0	$Q_{(0)}$
0	1	0
1	1	1

3.2.2.2 The D-Latch

A D-latch has an interface illustrated in Figure 3.19. This circuit has two inputs: D, the data input, and C, the control input. Although it has two outputs, it really only outputs Q, with another output of $\overline{Q}$ produced simply for the convenience of the user. The device supports two operations: *lock*, during which the device keeps the value stored, and *load*, during which a new value is loaded into the device. The C input controls the operation being performed. When C is 0, which is the normal case, the device is locked. When C is 1, the device loads a new value. The value loaded in is whatever is currently on the D line. The behavior of the D-latch is summarized in the *excitation table* shown in Table 3.6. The table shows the change in state for the latch, as a function of the operation specified by C.

Table 3.6 is split into two halves, in the same way that a truth table would be. In the left half is shown all possible values of the input parameters. The right half of the table shows the output of the circuit, Q. The table shows a change of state, using a subscripting system. In this scheme, $Q_{(0)}$ indicates the state of the latch before the operation, and $Q_{(1)}$ indicates the new state of the latch. The first line of the table indicates a lock operation, and shows that the state does not change. The two other rows in the table indicate a load operation, showing that the state of the latch changes to whatever value has been placed on the D line.

A useful way of demonstrating the behavior of a sequential circuit is through the use of a timing diagram. A timing diagram shows the behavior of the device over time, for a particular scenario. In the timing diagram scenario shown in Figure 3.20, the latch originally stores a 0, shown as the output Q. Initially the input C is held at 0, so that the latch keeps its value until the first rising edge. At this point the latch opens up, and allows the value D to enter. Since D has a 1 signal, the value stored in the latch becomes 1. The latch remains open until the falling edge of C, and so we see the value stored in the latch drop and rise as D changes. At the first falling edge, the latch slams closed, and the last value of D is trapped in the D-latch until it reopens at the next rising edge.

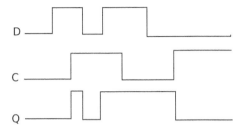

FIGURE 3.20 Timing diagram showing D-latch operation.

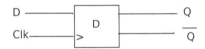

FIGURE 3.21 D-flip-flop interface.

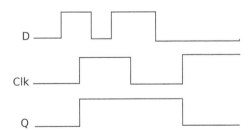

FIGURE 3.22 Timing diagram showing the D-flip-flop operation.

TABLE 3.7   D-flip-flop excitation table.

D	Clk	$Q_{(1)}$
X	⤴̸	$Q_{(0)}$
0	↑	0
1	↑	1

3.2.2.3   The D-Flip-Flop

We now turn our attention to the D-flip-flop. This is a clocked device, as already mentioned. The block diagram for the D-flip-flop is shown in Figure 3.21. The most noticeable difference between the D-latch and the D-flip-flop, is that the control input has been replaced by a clock input. In terms of differences in functionality, the D-flip-flop is only open at the trigger edge of the clock signal, as opposed to the D-latch, which is open as long as the C input is set. If we examine the same scenario given for the D-latch, it would result in the timing diagram in Figure 3.22. When the flip-flop opens at the trigger edge of the clock signal, the value of D, 1, goes into the device. It is stored there until the next rising edge of the clock, at which time the value of the D input has been lowered to 0. This new value is then stored in the flip-flop.

The behavior of the D-flip-flop is summarized in the excitation table marked Table 3.7. In this table, some novel notation is needed to handle the concept of the trigger edge. An arrow indicates that the trigger edge is detected by the flip-flop. and an arrow with a slash indicates that the flip-flop is not currently seeing the trigger edge on the clock input line.

The D-style device is not the only type of storage device available, although it turns out to be one of the most useful of the available devices. We will briefly cover one other type of storage device to demonstrate how flip-flop styles differ.

FIGURE 3.23   J-K-latch and J-K-flip-flop interfaces.

TABLE 3.8   J-K-Latch excitation table.

J	K	$Q_{(1)}$
0	0	$Q_{(0)}$
0	1	0
1	0	1
1	1	$\overline{Q_{(0)}}$

TABLE 3.9   J-K-flip-flop excitation table.

J	K	Clk	$Q_{(1)}$
X	X	⤒	$Q_{(0)}$
0	0	↑	$Q_{(0)}$
0	1	↑	0
1	0	↑	1
1	1	↑	$\overline{Q_{(0)}}$

3.2.2.4   The J-K- Storage Device

The J-K style storage device comes in both a latch, the J-K-latch, and a flip-flop, the J-K-flip-flop. These devices have no data input. Rather, two control inputs take the place of a data line. The block diagrams of the J-K-latch and the J-K-flip-flop are shown in Figure 3.23. These devices have a J control input, which can be referred to as the *set*-control, and a K control, which can be referred to as the *reset*-control. The devices are capable of performing the following operations:

- Lock: The device keeps its current value. This operation is specified with $J = K = 0$.

- Set: The value of the device changes to 1. This operation is specified with $J = 1, K = 0$.

- Reset: The value of the device changes to 0. This operation is specified with $J = 0, K = 1$.

- Compliment: The value of the device is toggled from 0 to 1, or from 1 to 0. This operation is specified with $J = K = 1$.

The operation of the J-K devices is summarized in the excitation tables of Tables 3.8 and 3.9.

3.2.2.5   Flip-Flops with Extra Pins

As a final note on flip-flops, the interface to a flip-flop can be a bit more complex than the simple interface presented. For example, the interface presented

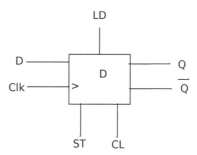

FIGURE 3.24 D-flip-flop with extra pins.

for the D-flip-flop had only four pins: D, Clk, Q, and $\overline{Q}$. In reality the interface might resemble more what is shown in Figure 3.24. This block diagram shows seven pins: the pins just listed, an ST, or *set* pin, a CL, or *clear* pin, and an LD, or *load* pin.

The ST and CL lines are *asynchronous*. The LD line is synchronous. A designation of asynchronous indicates that the operation being performed is done immediately, and without synchronization with the clock. In this way, the set and clear operations correspond closely to the operation of the J and K lines on the J-K-latch. When the ST line is asserted, the value stored in the flip-flop immediately changes to a 1, and when the CL line is asserted, the value stored in the flip-flop immediately drops down to 0. These operations are performed regardless of the clock signal. This is a useful feature. Often the flip-flops in a machine must have a specific initial value. The asynchronous lines can be used to load the flip-flop with either a 1 or a 0, even before the first clock trigger edge has been produced.

The LD line addresses one of the problems associated with the D-flip-flop. That problem is that the D-flip-flop can only store a value between clock pulses. It changes value every new clock cycle. The load line allows the flip-flop to ignore the clock. Ignoring the clock, the flip-flop does not see the trigger edge, and does not change its value, until the clock is re-enabled. Whether the flip-flop is locked or open to input is determined by the LD control line. When LD is a 1, the flip-flop is open, and any value on the D line will be latched by the flip-flop on the next trigger edge. If the LD line is 0, the flip-flop is locked, and no change will occur on the next trigger edge.

It is possible to construct a D-flip-flop, with a load line, from a D-flip-flop without a load line. This is demonstrated in the circuit schematic diagram

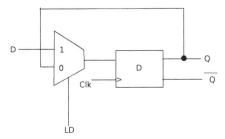

FIGURE 3.25 Implementing an LD pin on a standard D-flip-flop.

of Figure 3.25. Creating a flip-flop with load from one without load, is only mildly interesting. What is more interesting is the technique. In this technique, a multiplexer is used to choose the input to the flip-flop on every clock cycle. If the LD line is 0, then the old value of the flip-flop is fed back in as its new value. Although it might be said that the flip-flop is still changing value every clock cycle, the practical result is that the value stored in the device remains the same. When the LD line is set to 1, the value on the D line is sent into the device, causing the flip-flop to change its stored value. The structure used in this technique is called a *feed-back loop*, referring to the situation in which the old value is used in the computation of the new value.

3.2.3 Sequential Design

When we discussed combinational circuitry, we started our discussion by describing truth tables. These are tools that are used to give exact specification of the behavior of a combinational circuit. From this description, we were able to mechanically build Boolean expressions, giving the structure of the circuit, and finally implement the Boolean expression using gates.

We need a tool for sequential circuitry, that corresponds to the truth table. That is, we need a tool to specify the behavior of a sequential circuit. This tool must capture how the circuit moves from one state to another, and how the movement from one state to the next is affected by input, and how it affects output. The tool that is used most often is the *Finite State Machine* (FSM). Finite state machines are often also called *finite state automata* (FSA).

3.2.3.1 The FSM and State Diagrams

An FSM can be described as a collection of states, connected by transitions. The FSM produces output, and receives input, as well. Although a much more

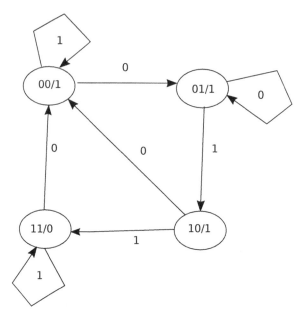

FIGURE 3.26 FSA0 state diagram.

formal definition of an FSM is possible, possibly a better introduction to FSMs would be through an example.

FSA0: You will find a graphical representation of an example FSA in Figure 3.26. These graphical representations are called *state transition diagrams*, or more succinctly, *state diagrams*.

Figure 3.26 shows four states. Each state is represented by an oval. The arrows are the transitions from one state to another. States are labeled with a two-part label of the form S/P, where S is a state number and P is an output value. For instance, the state in the lower right-hand corner has state number 10. This is a binary 2, and so we would refer to this state as State 2. The output associated with this state is 1.

Transitions also have labels. The label is of the form I, where I is the input associated with the transition. For example, the transition from State 1 to State 2 is taken if the current input is 1.

FIGURE 3.27 FSA0 interface.

The circuit being described by the state diagram in Figure 3.26 could correspond to the pin-out diagram of Figure 3.27. The block diagram shows a machine, FSA0, with a 1-bit input, i, and a 1-bit output p. The FSM of Figure 3.26 also shows that the circuit, internally, has four states. As with all sequential circuits, it moves from one state to another, each clock cycle. The state diagram describes how this is done. Let us read through the above state diagram to illustrate this. Starting in State 0, while in State 0 the output p should show a 1. At the next trigger edge, if the input i is a 1, the machine would remain in State 0. If, however, the input were 0, then the machine would move to State 1. In State 1, the output should be 1. An input of 0 causes the machine to remain in State 1, on the next trigger edge. An input of 1 causes the machine to change its state to State 2. In State 2, the output is 1; an input of 0 causes the state to change to State 0, and an input of 1 causes the state to change to State 3. Finally, in State 3, the output is 0; an input of 1 causes the state to remain unchanged, and an input of 0 causes a change of state to State 0.

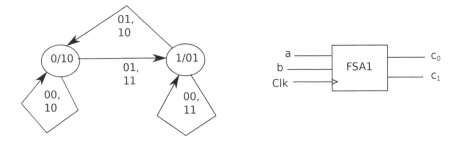

FIGURE 3.28 FSA1 state diagram and interface.

FSA1: Let's look at a second example, illustrated with the FSA and block diagram in Figure 3.28. The example FSA1 has only two states. There are two bits of input, a and b, and so the transitions are labeled with 2-bit configurations. The output is also 2-bit, c_1 and c_0, and so the states have 2-bit output labels. The diagram contains eight transitions, although you will only observe

TABLE 3.10   State transition table for
FSA0.

i	$Q_{(0)1}$	$Q_{(0)0}$	$Q_{(1)1}$	$Q_{(1)0}$	p
0	0	0	0	1	1
0	0	1	0	1	1
0	1	0	0	0	1
0	1	1	0	0	0
1	0	0	0	0	1
1	0	1	1	0	1
1	1	0	1	1	1
1	1	1	1	1	0

four arrows. To make the diagram less cluttered, arrows have been combined. For instance, the arrow from State 0, back to State 0 actually represents two arrows: an arrow labeled 00, and an arrow labeled 10. Notice an important property of transitions; all transitions from a particular state are mutually exclusive. That is to say that an input on one transition does not occur on any other transition emanating from the same state.

3.2.3.2 The FSM and the State Transition Table

FSAs have two commonly used representations. The first representation, which we have just finished examining, is the state diagram. This is a graphical representation of the state machine. The second representation is the *state transition table*, or just the *transition table*. This representation is textual. The two representations contain exactly the same information. However, when we are building circuits, a textual representation is necessary, in order to develop equations that can be implemented as digital circuitry.

FSA0: The transition table for the sequential circuit FSA0 is shown in Table 3.10. This table, like a truth table, has two halves. The columns on the left form the input half of the table, and the columns on the right form the output half of the table. The input half of the table starts with columns for all input variables for the circuit. These are then followed by columns for the current state of the machine. On the output side of the table, first we see columns for the next state of the machine, followed by columns for the circuit output variables. Notice that when building the table, the clock signal is not considered an input, and does not have its own column. In fact, since all sequential circuits are clocked, the clock input is usually taken for granted, and ignored.

As with the state diagram, the transition table shows two things: the output for each state, and the movement from one state to another, given a particular input. The output p is based on the current state, called $Q_{(}0)$, and composed

TABLE 3.11   State transition
table for FSA1.

a	b	$Q_{(0)}$	$Q_{(1)}$	c_1	c_0
0	0	0	0	1	0
0	0	1	1	0	1
0	1	0	1	1	0
0	1	1	0	0	1
1	0	0	0	1	0
1	0	1	0	0	1
1	1	0	1	1	0
1	1	1	1	0	1

of the two bits, $Q_{(0)1}$, the high-order bit, and $Q_{(0)0}$, the low-order bit. You can, for instance, verify that, just as the diagram shows an output of 1 for State 0, the first row of the table shows an output $p = 1$, for the state $Q_{(0)} = 00$. Similarly, the state diagram shows that in State 3 the output is 0, and in the transition table, the fourth and eighth row show an output of $p = 0$ for state $Q_{(0)} = 11$. You can also observe that the output column shows the same four outputs repeated twice. This is because the states are repeated twice: first with input 0 and then with input 1.

Movement from one state to another is shown, in the table, using the input columns, the current state columns, and the next state columns. In our example, the next state is denoted by $Q_{(1)}$, which is composed of the two bits $Q_{(1)1}$ and $Q_{(1)0}$, which represent the high-order and low-order bits of a binary number, respectively. As an example, observe that the state diagram in Figure 3.28 shows a transition from State 2 to State 0 on input 0. The same transition can be seen in the third row of the transition table, where input $i = 0$ causes the current state $Q_{(0)} = 10$ to change to the next state $Q_{(1)} = 00$. You can, in fact, verify that each arrow in the state diagram becomes a single row on the transition table.

FSA1: We give the transition table for the example FSA1 in Table 3.11. The only interesting difference in this table, from Table 3.10, is that there are now two inputs, and so two columns for input in the transition table, and two outputs, with corresponding columns in the table.

3.2.3.3   State Diagrams and Transition Tables: Building One Representation from the Other

As stated, the state diagram and the transition table contain the same information. Constructing one from the other is straightforward. To construct a diagram from a table, you would do the following:

- Lay down states using numbers from the current state column.

- Fill in outputs from the output columns.

- Draw arrows, one per row in the state table, from the current state to the next state.

- Fill in the input labels on the diagram, from the input columns in the table.

Going from a state diagram to a transition table you would reverse the procedure. In particular you would do the following:

- Create the state table heading, listing out the input variables, the bits of the current state number, the bits of the next state number, and the output variables.

- Fill in all possible bit configurations on the input half of the table.

- On each row, fill in the output for the current state.

- On each row, fill in the next state, using the arrow in the state diagram corresponding to the row in the transition table.

One problem with this second procedure is determining the number of bits in the state number. However, determining the number of bits in the state number is a mechanical procedure. If your machine has m states, then the state number will have $\lceil \log m \rceil$ bits. For example, if you have 4 states, then there will be $\lceil \log 4 \rceil = 2$ bits, with current state names $\langle Q_{(0)1} \, Q_{(0)0} \rangle$, and next state names $\langle Q_{(1)1} \, Q_{(1)0} \rangle$.[1]

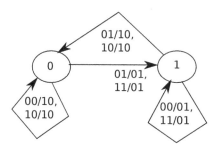

FIGURE 3.29 Mealy machine for FSA1.

---

[1]The notation $\langle xy \rangle$ indicates the concatenation of the value of x and the value of y.

3.2.3.4 Moore versus Mealy Machines

At this point, we have presented a fairly clean description of FSMs. However, there is a small complexity which needs to be mentioned. That complexity is that there are actually two different types of FSMs: the *Moore machine*, and the *Mealy machine*. The difference between the two is with what the output is associated. On the Moore machine, output is associated with the state. This is the type of machine we have been using so far. On the Mealy machine, the output is associated with the transition. In Figure 3.29, we give the Mealy machine for the example FSA1. As you can see, the state label now only gives the state number, S, whereas the transition label now has two parts, I/P giving the input value, and the output value, in that order. This small change means that the output is no longer determined by the current state, but rather a combination of current state and input. That is to say, the output is determined by the next state.

We do not discuss the Mealy machine in detail here. The Mealy machine is useful in some situations, but the Moore machine turns out to be a little easier to use in circuit design. For this reason, we will deal only with the Moore machine in our discussion.[2]

3.2.3.5 Implementing a Sequential Design

We now have a tool for specifying the behavior of sequential circuits: the FSM. An FSM can be described as either a state diagram, or as a transition table. The next step is to turn the FSM into circuitry. This process is similar to the process used when turning a truth table into circuitry. It consists of copying out equations from the table, for all variables on the output side of the transition table.

All sequential circuits are implemented using the same structure. An abstract diagram of this structure is found in Figure 3.30. The structure consists of two major blocks: a *register* and *control* circuitry. The register is a collection of flip-flops, one per bit in the state number, that is used to store the current state. In the FSA0 example you would have two flip-flops: one to store $Q_{(0)1}$, and one to store $Q_{(0)0}$. For FSA1 you would have one flip-flop, storing $Q_{(0)}$. All flip-flops in the register are connected to the clock.

The control circuit is a combinational circuit. As can be seen, it takes as input the current state, and the inputs to the circuit. It calculates, as output,

---

[2]We are not suggesting that the Mealy machine be dismissed. Mealy machines can often produce circuitry that is more efficient than Moore machines, although the two machine types have been proven equivalent.

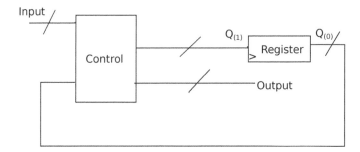

FIGURE 3.30 Sequential circuit structure.

the outputs of the circuit, and the next state. The next state is then loaded
into the register on the next clock cycle, causing the machine to transition.

FSA0: The most interesting part of converting an FSA into the circuit
of Figure 3.30 is the construction of the control circuitry. We demonstrate
the process with the FSA0 example. The process is already largely developed, since we are simply implementing a combinational circuit. The process
of implementation begins by copying out equations for each variable in the
output side of the transition table, Table 3.10. These can then be simplified
in Karnaugh-maps. When this is done we get the following equations.

$$Q_{(1)1} = iQ_{(0)0} + iQ_{(0)1} = i(Q_{(0)0} + Q_{(0)1}) \tag{3.13}$$

$$Q_{(1)0} = \bar{i} \cdot \overline{Q_{(0)1}} + iQ_{(0)1} + Q_{(0)1}\overline{Q_{(0)0}} \tag{3.14}$$

$$p = \overline{Q_{(0)1}} + \overline{Q_{(0)0}} \tag{3.15}$$

Equations 3.13 and 3.14 give the two bits of a new state number, given the
old state number, and current value of the input. This is fed into the two flip-
flops of the state register, and when the clock trigger edge unlocks the state
register flip-flops, the register will be loaded with the new state. On each clock
cycle, the state register will be loaded with the new state, causing the machine
to transition through a sequence of states, as specified in the state diagram in
Figure 3.26.

The control circuit also calculates the output of the machine, using Equation
3.15. This output is calculated using the bits of the current state, and ensures
that the output corresponds to the current state, as specified in Figure 3.26.

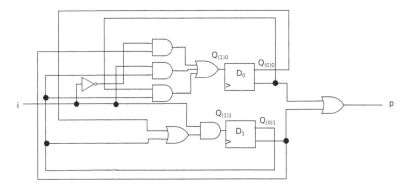

FIGURE 3.31 Schematic of the FSA0 circuit.

We are now ready to build the circuit. We first build the register out of two D-flip-flops: one flip-flop to store the low-order bit of the state number, and one to store the high-order bit. We call these flip-flops D_0 and D_1, respectively. Next, we implement the equations with gates. The result for FSA0 is shown in Figure 3.31. Notice that the D input to each flip-flop is the next state of the flip-flop, and the Q output is the current state of the flip-flop. All of the gates implementing the above equations are considered components of the control circuit.

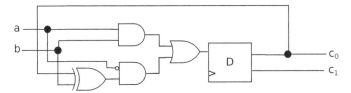

FIGURE 3.32 Schematic of the FSA1 circuit.

FSA1: We now examine FSA1. It is implemented in the same way as FSA0. The equations copied from the transition table, via a K-map, are as follows.

$$Q_{(1)} = ab + \overline{a}\overline{Q_{(0)}}b + \overline{a}Q_{(0)}\overline{b} = ab + \overline{a}(Q_{(0)} \oplus b) \qquad (3.16)$$

$$c_1 = \overline{Q_{(0)}} \qquad (3.17)$$

$$c_0 = Q_{(0)} \qquad (3.18)$$

The circuit will have a one flip-flop register, and is shown in Figure 3.32.

Notice that in the equation for $Q_{(1)}$ we have done a little algebraic simplification. This allows us to combine two single groups, that came out of the Karnaugh-map, using the XOR function.[3]

3.2.4 Sequential Circuit Analysis

We now turn our focus to a new topic. At this point we should be able to design a sequential circuit, using an FSM as a circuit specification. The opposite of design is analysis. Design is the procedure of implementing a circuit from a specification. For example, we would be given a state diagram, perhaps, and be asked to produce a circuit schematic. The opposite of this, analysis, is the procedure we would follow to produce a state diagram from a circuit diagram. Analysis is something we would be required to do less frequently, in processor design, but it is worth, at least, demonstrating how this is done.

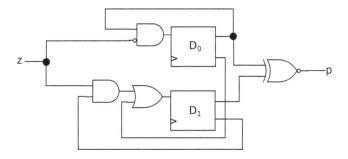

FIGURE 3.33 Analysis example sequential circuit.

Let's look at an example. In Figure 3.33 is a circuit diagram of a sequential circuit. We start the analysis process by copying out equations. We need equations for all circuit outputs, in this case just p, and all flip-flop inputs, in this case $Q_{(1)1}$ and $Q_{(1)0}$. These equations are derived by tracing the circuitry, determining which gates are connected to the line of interest, and what is connected to the input of the gates. In our example, we might start with the output p. We observe that p is connected to an XNOR gate. The inputs of the XNOR gate are $Q_{(0)0}$ and $Q_{(0)1}$. So, the equation for p would be

$$p = Q_{(0)0} \odot Q_{(0)1}. \tag{3.19}$$

---

[3]Remember that the XOR function results in a K-map with a checkerboard pattern. Often, if you observe a partial checkerboard pattern in a K-map, you will be able to combine singleton groups using an application of the XOR, or XNOR function.

TABLE 3.12   State transition table for
Figure 3.33 example.

z	$Q_{(0)1}$	$Q_{(0)0}$	$Q_{(1)1}$	$Q_{(1)0}$	p
0	0	0	1	0	1
0	0	1	0	1	0
0	1	0	1	0	0
0	1	1	0	1	1
1	0	0	1	0	1
1	0	1	0	0	0
1	1	0	1	0	0
1	1	1	1	0	1

In a similar way we trace the input to the D_0 flip-flop and the D_1 flip-flop, and arrive at the following equations for $Q_{(1)0}$ and $Q_{(1)1}$, respectively.

$$Q_{(1)0} = \overline{z}Q_{(0)0} \tag{3.20}$$

$$Q_{(1)1} = \overline{Q_{(0)0}} + zQ_{(0)1} \tag{3.21}$$

We are now ready to build the transition table. This is done in the same way that we would build a truth table from equations. That is to say, we lay out the table, fill in all possible input values, and then calculate the output values, using the equations, possibly with the aid of intermediate columns. The resulting table, for our example, is shown in Table 3.12.

As the last step, we can now construct the state diagram from the transition table. This process was previously covered. The result is the diagram in Figure 3.34.

3.2.5 Common Sequential Circuits

The basics of sequential circuitry have now been covered. In this section we present several sequential circuits that are particularly useful in the design of a processor. The circuits that are most useful can be classified as *registers*.

A register is a storage device. Unlike the flip-flop, or the latch, it is capable of storing multi-bit data. So, for example we might wish to store 4-bit numbers. An example of a 4-bit number might be 1100, which is the binary representation of the decimal number 12. This number can easily be stored by storing each of the four bits in a separate flip-flop.

Each bit in a multi-bit number is, by convention, given an index. You start with the bit furthest to the right, which is the low-order bit. The index of this

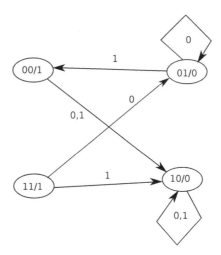

FIGURE 3.34 FSA for the Figure 3.33 example.

bit is always 0. The indexes increase as you go further left, to higher-order bits. In a 4-bit number x, we would have the string $\langle x_3\ x_2\ x_1\ x_0 \rangle$.

In this section we present three types of useful registers. Although registers all store values, the type of a register refers to additional operations the register performs. We begin with what we often call the *simple register*. The more descriptive name for this register is the *parallel-load register*.

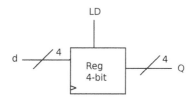

FIGURE 3.35 Parallel-load register interface.

3.2.5.1 The Parallel-Load Register

A parallel-load register performs two operations: *lock* and *load*. The operation is specified by a load control line, called LD. The interface for the simple register is shown in Figure 3.35. The register operates very much like a flip-flop with a load line. That is to say, when LD is 0 the register is locked,

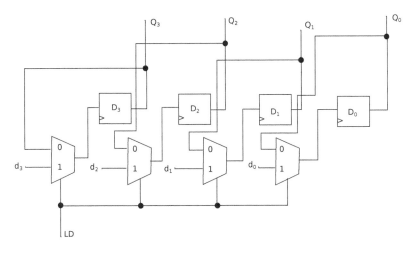

FIGURE 3.36 Schematic for a parallel-load register.

and when LD is 1 the register latches the d value on the next clock cycle. This register could be designed in the same way as we have been doing with custom sequential circuits; we could convert an FSA into a circuit. A more convenient way to design it is using the same method we used to design the D-flip-flop with load. We would then use flip-flops and multiplexers, resulting in the diagram of Figure 3.36, for the 4-bit simple register. Each flip-flop, in this diagram, is marked with the index of the bit it stores. Each flip-flop is set up with a feedback loop, and a multiplexer that chooses between the current value of the flip-flop, or the new bit from the input d. The LD line controls all multiplexers causing them to perform the same operation, and the clock inputs on the flip-flops are all connected to the processor clock, causing all changes to be performed synchronously.

3.2.5.2 The Shift Register

Our next register is the *shift register*. This register also has two operations. The load operation is now replaced with a shift operation, yielding a register with operations *shift* and *lock*. It has a single control line that controls the operations called SH. When SH is a 0, the register is locked, and when SH is a 1 the register performs a shift operation on the next clock cycle.

In a shift operation, each bit of the register is moved over to the next position. A shift can be either to the left or to the right. Both the left and right shift are illustrated in Figure 3.37. In the left shift, each bit is copied over to the next bit to the left. This means that the low-order, or rightmost

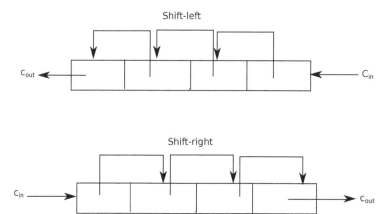

FIGURE 3.37 Illustration of shifting.

bit is vacated. An input called the *carry-in* is placed into the vacated spot. Also, the leftmost, or high-order bit falls off the end of the register. This bit appears as an output called the *carry-out*.

As an example we will design a left-shift register. The interface and schematic are shown in Figure 3.38. Notice that we use the same design technique as with the simple register; that is, a multiplexer to implement multiple operations. To perform the shift, we simply feed in the value of the next bit to the right as the new flip-flop value. So, the current value of each flip-flop is fed back into itself, as Option 0 on the multiplexer, and into the next flip-flop as Option 1 on the multiplexer. c_{out} is just the value of the leftmost flip-flop, and c_{in} is fed into the rightmost flip-flop.

3.2.5.3 The Counter

The last register we present is the counter. A counter has two operations: the usual lock operation and the *increment* operation. The control line IN is set to perform an increment, and cleared to lock the register. An increment operation adds 1 to the contents of the register on the next clock cycle. If you watch a 4-bit counter during successive increment operations, you will see its contents go through the sequence 0000, 0001, 0010, 0011, 0100, 0101, 0110, 0111, 1000, 1001, 1010, 1011, 1100, 1101, 1110, 1111, and on the next clock cycle the counter cycles back to 0000, and repeats the cycle.

To construct a counter we take the same simple register we have presented, and replace one multiplexer input with a connection to circuitry that incre-

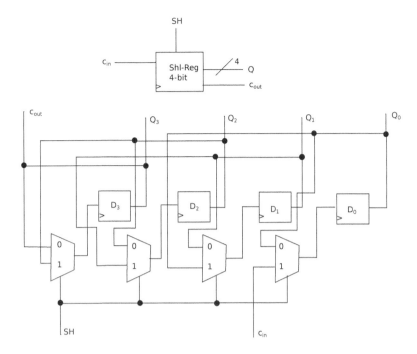

FIGURE 3.38 Interface and schematic for a shift register.

ments the flip-flops. The interface and the schematic for the counter are shown in Figure 3.39. The full adders in the diagram form a 4-bit ripple-carry adder that adds 1 to the low-order bit, and propagates the carries up through the bits to the left. The counter carry-out is the carry-out from the high-order adder. This line produces a signal of 1 when the counter rolls over from fifteen to zero.

3.2.5.4 The Standard Register

Before we finish this chapter, it is worth mentioning that we might build registers that combine the three registers we have discussed, or combine them with other types of registers. In subsequent discussion we make use of what we call the *standard register*. This register performs four different operations: the usual lock operation, the load operation that we discussed, the increment operation just presented, and a *clear* operation, in which the contents of the register are reset to zero on the next clock cycle. The register is controlled by the lines LD, IN, and CL. It is a combination of the simple register, the counter, and a clearing register, which we have not discussed. The pin-out diagram for the standard register, and its schematic are shown in Figure 3.40.

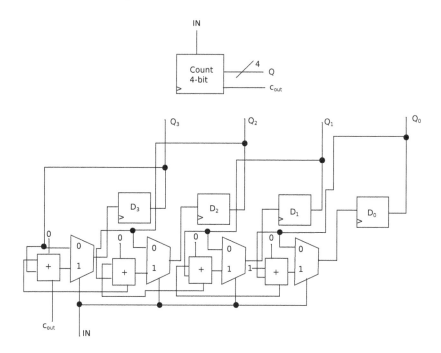

FIGURE 3.39 Interface and schematic for a counter.

The diagram uses some abbreviated notation for structures we have already used. First, we see a D-flip-flop that has a 4-bit input and 4-bit output. This is nothing more than a succinct notation for the 4-flip-flop array we have been using to draw all of our registers. We also see a multiplexer that has 4-bit inputs and output. The size of this MUX is 4-4, indicating that it has 4 inputs, and that each input is 4 bits wide. Again, this is nothing more than the array of 4 single-bit multiplexers used in the previous diagrams. Each of the four multiplexers feeds one of the four flip-flops. The same goes for the 4-bit adder we use, which is a collection of four full adders, arranged as a ripple-carry adder.

The circuit in Figure 3.40 is controlled by an encoder. Four trigger lines are input to the encoder: the LD line, the IN line, the CL line, and a line calculated using a NOR gate, that produces a 1 signal when all of the other three control lines are 0. This encoder causes the multiplexer to allow through Option 0, the current value of the flip-flop array, if LD, IN, and CL are all 0; Option 1, a 4-bit 0 value, if CL is a 1; Option 2, the current value of the flip-flop array incremented, if IN is a 1; and Option 3, the new value of the flip-flop array, d, if LD is a 1.

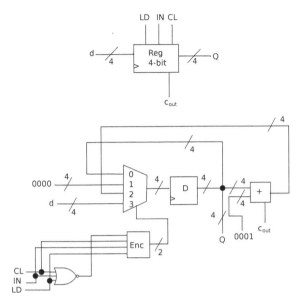

FIGURE 3.40 Interface and schematic for a standard register.

3.3 SUMMARY

In this chapter we have explained how digital circuits function, and how they are designed. We discussed the difference between combinational circuits and sequential circuits. It was pointed out that a combinational circuit has no memory, and a sequential circuit does.

The design process for both sequential and combinational circuits was described. This process takes a specification in tabular form, derives equations from the specification, and implements them using digital components.

The concept of clocking was described. The low-level devices used in a computer were also discussed. And, at this point we are ready to start looking at larger devices.

3.4 EXERCISES

3.1 Draw circuit diagrams for the following functions.

a. $\overline{a + b \cdot (c \oplus b)}$

   b. $(a+b)(\bar{a}+\bar{b}) \odot (a \oplus b)$

   c. $(\bar{c} + (\bar{b} + \bar{a})(b + c))a$

3.2 For the following function:

$$h = \overline{x}\overline{y}\overline{z}\cdot\overline{w} + xy\overline{z}w + \overline{x}yzw + xyz\overline{w} + \overline{x}\cdot\overline{y}\cdot\overline{z}\cdot\overline{w} + x\overline{y}z\overline{w} + \overline{x}\cdot\overline{y}zw + x\overline{y}\cdot\overline{z}w$$

   a. Draw the truth table for the function h.

   b. Show the Karnaugh-map for h.

   c. Give the simplified equation for h, derived from the K-map.

   d. If possible, further simplify your answer to Part c using algebraic manipulation.

   e. Draw the circuit schematic for the simplified circuit.

3.3 You are to design a divider circuit. Your circuit takes in a 3-bit number, x, and outputs a 2-bit number p. The output p is defined as $p = \lfloor x/3 \rfloor$. Show your truth table, K-maps, simplified equations, and circuit diagram.

3.4 It is possible to extend the MUX tournament composition method from Section 3.1.2.4 to build multiplexers with more inputs. This is done by either using larger multiplexers for the heats, or including more rounds in the tournament.

   a. Draw a diagram showing how to build an 8-1 MUX, using two 4-1 MUXs and a 2-1 MUX.

   b. Draw a diagram showing how you would build an 8-1 MUX using four 2-1 MUXs and a 4-1 MUX.

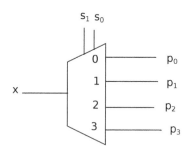

FIGURE 3.41 Interface for a DMUX.

3.5 A priority encoder is an encoder for which you can set several inputs simultaneously. The code that is produced, in our version, is the code of the input line with the highest index. For example, if the encoder shown in Figure 3.8 were a priority encoder, and if input lines p_2 and p_3 were both set, and the rest of the input lines were clear, the code produced would be 3 (11 in binary). Design the 4-2 encoder, showing the truth table, Karnaugh-maps, simplified equations, and circuit diagram.

3.6 Redesign the 4-1 MUX using no other components than the tri-state switch and a decoder.

3.7 A *half adder* is a circuit with two inputs, a and b, and two outputs, s, the sum bit, and C_{out}, the carry-out. This differs from the *full adder* that was presented in Section 3.1.2.5, in that the full adder had three inputs. But like the full adder, the half adder adds its inputs producing the sum and carry. Design the half adder. Show a truth table, any Karnaugh-maps, simplified equations, and a circuit diagram.

3.8 A T-flip-flop is a flip-flop with one input, T. When T is a 1, the flip-flop complements its contents on the next clock cycle. When T is 0, the flip-flop is locked. Design a T-flip-flop using

   a. a D-flip-flop and any extra gates required or

   b. a J-K-flip-flop and any extra gates required.

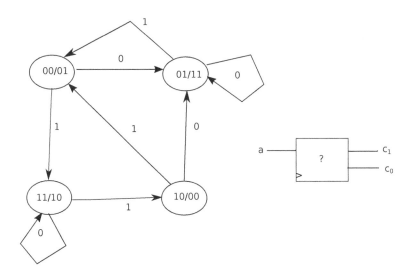

FIGURE 3.42 FSM and interface for Exercise 3.9.

3.9 Design the circuit specified by the FSM in Figure 3.42. Show the transition table, any K-maps used, simplified equations, and a circuit diagram.

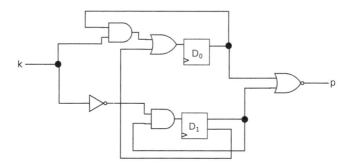

FIGURE 3.43 Schematic for Exercise 3.10.

3.10 Analyze the circuit drawing of Figure 3.43. Show the equations, the transition table, and the state diagram for the circuit.

3.11 Gray code is the numbering system used to fill in the coordinates of a K-map. You are to build a 2-bit Gray code counter that counts the sequence 00, 01, 11, 10, 00, 01, 11, 10, ⋯. Your counter will have one input, IN, that causes the counter to increment on the next clock cycle, if set, and causes the counter to lock, if clear. Design your circuit as an FSA. Draw a state diagram for the counter, a transition table, give simplified equations for the counter, and draw the circuit diagram.

3.12 Design a 4-bit shift register with a single control line, L. When $L = 1$, the shift register shifts left on each clock cycle. When $L = 0$, the register shifts right. There is no lock operation. Draw a diagram showing flip-flops and MUX connections.

Devices and the Bus

CONTENTS

I N THE LAST CHAPTER, we explored the low-level circuitry in the processor. Remember that we wish to build a processor, and so the discussion in the last chapter lays the groundwork for that exercise. However, to build a processor, we must still develop an understanding of the environment in which a processor works. This chapter looks at devices connected to the processor, and how the processor interacts with them.

Most of the devices with which the processor interacts are on the *motherboard*, a term used to refer to the circuit board on which the processor is mounted. Most of the devices with which the processor directly communicates can be classified as either memory devices or peripheral devices. Collectively, we refer to all of the devices with which the CPU communicates as *external devices*, which emphasizes that they are external to the processor.

The processor, or *central processing unit* (CPU), needs to be able to communicate with all of these other devices. And so, all of the devices in the computer system are connected by wires, which are used to send signals from one device to another. There are several options in terms of connection topology. One option is to connect every device with every other device with which it needs to communicate. We often refer to this strategy as the *direct connection* strategy. The direct connection strategy has an advantage. It allows concurrent communication. That is, while one pair of devices are communicating, another pair can also be communicating on a separate connection. But, the direct connection strategy also has a disadvantage; there is a lot of wiring, which increases the complexity of the circuitry.

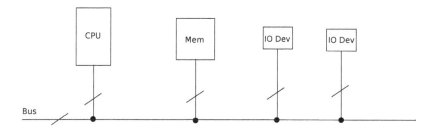

FIGURE 4.1 A bus connection.

A second option is referred to as the *bus connection* strategy. It uses a single connection path, called a bus. This strategy is illustrated in Figure 4.1. The figure shows several devices connected to the bus. These include the CPU, a memory unit, and a couple of peripheral devices.

The Bus connection strategy has the advantage of creating a simple communication topology. The disadvantage is that, now, only a single pair of devices can communicate at a particular time. If another pair needs to communicate, that pair must wait until the bus is released.[1] This problem, however, is not

---

[1] It is possible to ameliorate the problem of competition for the bus by building a system with multiple buses. However, the more buses in your system, the more the system begins to resemble the direct connection system, and just as with direct connection, circuit complexity becomes an issue.

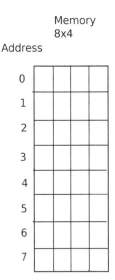

FIGURE 4.2 Conceptual drawing of a memory unit.

all that debilitating. In computer systems, most communication is instigated by the CPU. As a result, bus request conflicts are unusual.

To more fully understand the communication between devices, we need to examine the different types of devices in more detail. This is done in the next few sections.

4.1 MEMORY

It is almost impossible to do computation without some kind of device for storage of data and programs. We have already discussed one such device: the register. This device is inside the processor. Because we wish to limit the size of the processor, to increase its speed, only a small number of registers can be included in the CPU. This limitation, then leads to a search for a way of storing data outside the CPU, and is the reason that most computer systems entail some kind of external memory device.

4.1.1 Memory Operation

Let's begin our discussion of memory by considering what a memory unit does. A memory unit is a storage device. It stores multi-bit values, but unlike a register, it is capable of storing many multi-bit values. A conceptual illustration of a memory unit is shown in Figure 4.2.

A memory unit is a collection of storage devices called *words*. Each word is capable of storing a multi-bit value. In our example, each word is a row in the diagram of Figure 4.2, composed of four bits. The memory unit contains eight such words. The number of bits in a word is called the *width* of the memory unit, and the number of words the unit contains is called its *length*. The *size* of the memory unit is usually given by specifying its length and width. So, in our example, the size of the memory unit would be 8×4.

We would like to be able to perform operations on the memory unit. For example, we might want to store a value of 5 (0101 in four bit notation) in the word on the third row. To do this we need a way of identifying the words in the memory unit. We do this by sequentially numbering each word in the unit. The index of a word is called its word *address*. From Figure 4.2 we see that the address of the word on the third row is 2. Our request to the memory unit would then be to write a 5 at memory location 2.

Memory units, for the most part, are capable of performing two operations:

- *Read*: produce the contents of a particular memory location.

- *Write*: store a given value in a particular memory location.

We have already given an example of a write operation. A read example might be a request that the value stored at address 5 be produced for examination.

4.1.2 Memory Types: ROM and RAM

There are many ways of classifying memory units. One way to classify them is as

- Read Only Memory (ROM) or

- Random Access Memory (RAM).

The name for the ROM unit is pretty self-explanatory. These units only perform a read operation. This, of course, means that it is impossible to store any value in the unit, which would, initially, appear to render them useless. These units, however, can be manufactured with pre-installed content. This is useful; often computer manufacturers wish to give information to the purchaser of a computer; information used by the computer, but never changed by the user. As an example, there are low-level operating system components, like device drivers, that might be included in your computer. These programs are repeatedly executed, but never changed.

It is less obvious from the name what RAM does. RAM, however, might better be called Read Write Memory (RWM), but for historic reasons it received its actual name.[2] A RAM unit is capable of both read and write operations. This is the computer memory we normally think of as the working memory in the computer.

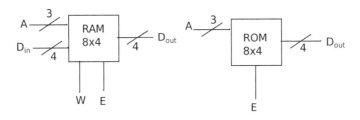

FIGURE 4.3 Interface for ROM and RAM.

Simplified pin-out diagrams for the two memory types are shown in Figure 4.3. These diagrams show the interface for 8×4 units. The RAM chip has two input ports, labeled D_{in}, and A. The D_{in} port allows data to be given to the RAM, when performing a write operation. The A port is the address port, allowing an address to be specified for both read and write operations. The RAM also has an output port, labeled D_{out}. This port produces the data at the specified location, during a read operation. Two control lines are also shown in the diagram, labeled W and E. The W line specifies the operation being performed. When $W = 0$, a read operation is under way, and when $W = 1$, a write operation is being performed. The E line is an enable line, that controls the output, D_{out}. When $E = 0$, the output lines show Z (high impedance). When $E = 1$, the contents of the specified memory location would show on D_{out}. You can accurately think of E as controlling tri-state switches on the output port D_{out}, that either stop output, or let it through.

The ROM unit in Figure 4.3 is similar to the RAM unit. But, because it only supports the read operation, it has no need for the D_{in} port. Likewise, with only one operation, it does not need a W control line to choose operations. It has only the E control input to control the output, D_{out}.

---

[2]The term Random Access Memory refers to the fact that, up to a certain point in time, computer memory was tape. When accessing tape, accessing a location at the beginning of the tape took less time than accessing a location at the end of the tape. With RAM, one could randomly, or arbitrarily, access any location, and it always took the same amount of time.

Let us summarize the operation of the memory units. To perform a write operation would require the following actions.

1. Set up the inputs.

   (a) Assert the desired address on the A port.

   (b) Assert the desired data on the D_{in} port.

2. Perform the operation by strobing (setting and then resetting) the W line.

Notice that the write operation can be performed in two steps if we assert both the address and the data simultaneously. This concurrency allows the processor to do the operation in two steps instead of three, speeding up the access operation. The read operation would be performed as follows.

1. Assert the desired address on the A port.

2. Strobe the E line, and allow time for the data to present itself on the D_{out} port.

Again, this operation takes two steps.

The memory units in Figure 4.3 are, as stated, 8×4. Memory units come in different sizes. In general, the size of a memory unit is $2^k \times m$. That is to say, their length is a power of two. Given the size of a memory unit, we can calculate the sizes of the data and address ports. The data ports, D_{in} and D_{out} would be of width m. The address port must be wide enough so that addresses between 0 and 2^k can be represented. This requires $\log 2^k = k$ bits. In the examples of Figure 4.3, the unit length was $8 = 2^3$, so $k = 3$, and so we needed a 3-bit address port.

4.1.3 Memory Composition

When we discussed multiplexers, we learned how to build larger multiplexers from smaller multiplexers. We might be interested in doing the same thing with memory units. The process of implementing a larger unit from smaller units is called *memory composition*. We often split composition into the categories of *horizontal composition* and *vertical composition*. We will introduce these two types of composition with a couple of examples.

4.1.3.1 Horizontal Composition

For horizontal composition, suppose that we have a pile of 8×2 RAM units. We need an 8×4 RAM unit. Figure 4.4 shows the details of this example. On the top-left we have a representation of the 8×4 RAM as an array of bits. To its right is the same RAM unit split in half. The strategy is to use two of

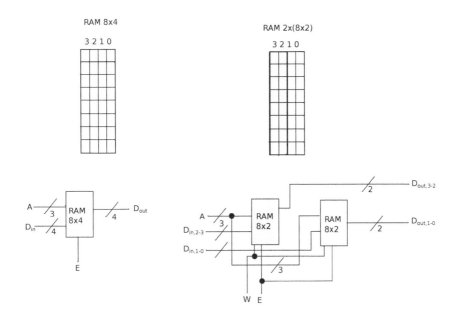

FIGURE 4.4 Horizontal composition.

the smaller RAM units to implement the large unit: one unit, filled with gray, stores the higher half of the word, and the other, unfilled, stores the lower half of the word. At the bottom of Figure 4.4, on the left, we show the interface for the 8×4 unit, and on the right we show how the 8×2 units are combined to create the larger RAM. The small units are stacked, horizontally, one next to the other. The same address and control lines are fed into both RAM units, causing them to issue out data from the same word, or store data to the same word. The bits of the input data are split between the two units. On the output side, the bits from the high-order RAM and the low-order RAM are assembled into the full 4-bit word.

4.1.3.2 Vertical Composition

For vertical composition, suppose that we have a pile of 2×4 ROM units, and we require an 8×4 ROM. The details of the example are shown in Figure 4.5. On the top-left, again, we see the conceptual diagram of an 8×4 ROM. On the right we show how we are implementing the larger ROM, using 2×4 RAMs. We use four such ROMs, each storing 2 words, and each marked in a different shade of gray. At the bottom, on the left, is the interface for the 8×4 ROM. On the right we see how to combine the smaller ROM units to implement the larger unit. A decoder is used to activate exactly one ROM

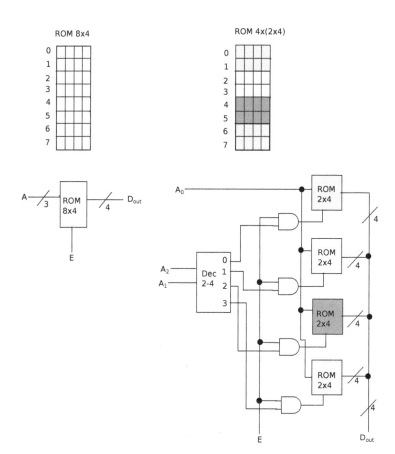

FIGURE 4.5 Vertical composition.

unit, when the E line is set. Output of that unit is released onto the D_{out} output bus.

In the example of Figure 4.5 we can think of splitting the address of the 8×4 ROM into two fields: the *internal address* and the *unit number*. In the diagrams there are four ROM units being used. We can give each of these units an index, with small indexes given to the ROM units storing values in low memory, and large indexes given to ROM units storing values in high memory. So, ROM Unit 0 stores words 0 and 1, ROM Unit 1 stores Words 2 and 3, ROM unit 2 stores Words 4 and 5, and ROM Unit 3 stores Words 6 and 7 of the larger unit. These unit numbers can all be represented with 2 bits in binary. We therefore use two bits from the address as the unit number. The bits we use are the high-order bits, as shown in Figure 4.6.

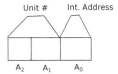

FIGURE 4.6 Address fields.

The 2×4 units each contain two words: Word 0 and Word 1. They can be addressed using a 1-bit address. The remaining bit of the 8×4 address is then designated as the internal address. The internal address is fed directly into each ROM in the diagram of Figure 4.5. Each ROM selects either its odd or even location. The decoder then uses the unit number to enable the appropriate ROM, and that ROM presents its output to D_{out}.

This technique for vertical composition is called *high-order interleaving*. The name refers to the fact that the high-order bits of the address are used as the unit number. In another scheme, called *low-order interleaving*, the low-order bits are used as the unit number. In our example, with low-order interleaving A_{1-0} would be used as the unit number, and A_2 would be the internal address.

What is important about the example in Figure 4.5 is the technique used in high-order interleaving. In general the full address, internal address, and unit number may be different sizes. But still, the procedure is to split the full address into two fields, wire the internal address to the address ports of the smaller memory units, and use the unit number to drive the address decoder.

4.1.4 Internal Memory Structure

Once we understand the interface for a memory unit, we might start wondering how the circuitry in the memory unit functions. The best way to understand this is to look at an example. We will construct a 4×2 RAM. A schematic diagram for this circuit is shown in Figure 4.7. The A port for this RAM is 2 bits wide, and the data ports are also 2 bits wide.

The RAM unit in Figure 4.7 uses 2-bit registers to store the data. The registers are arranged from top to bottom, address 0 to address 3. A decoder, the *address decoder*, is used to transform the address into trigger lines. These trigger lines are used to activate the appropriate register. Load lines on each register are connected to AND gates. Each AND gate checks whether that register has been selected by the decoder, and if a write operation has been specified. If so, the register is opened up to receive its data input. The input

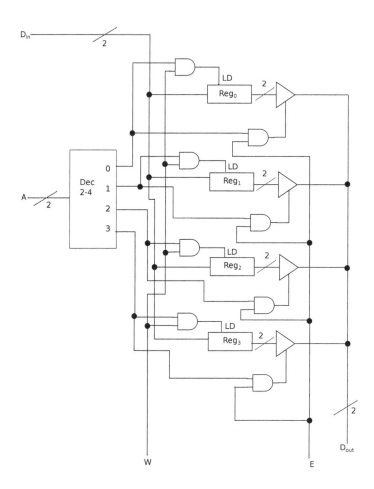

FIGURE 4.7 Structure of RAM.

port on each register is connected to the D_{in} input of the RAM. That is to say, the D_{in} input is waiting to enter all registers. The trigger line opens one register, for a write operation, and the data can enter that register only.

In Figure 4.7, a set of tri-state switches is used to perform the read operation. The output port of each register is connected to a tri-state switch, which outputs to the D_{out} port of the RAM. The switch is activated by another AND gate. This one checks to see if the register has been selected by the address decoder, and the output has been enabled. So, all registers are waiting to connect to the D_{out} port. All outputs are stopped by the switches, except the selected register, which connects to D_{out} when the enable line is activated.

As a note on our use of registers to implement RAM, the devices used in the construction of RAM are not clocked. They use latches, as opposed to flip-flops. This allows the RAM to function without a clock signal. It also speeds up the operation of the RAM, with operations performed instantly, rather than after waiting for the next clock pulse.

The structure of a ROM unit is similar to that of the RAM. The ROM unit would be missing the load circuitry, but would contain the tri-state switches for the read operation.

4.1.5   RAM Types

We have classified memory units already, by whether or not they are writable. We now look at further ways of classifying memory units. We start with a classification of RAM units, based on storage technology, and then consider a classification of ROM units, based on write-ability.

RAM units can be classified as follows.

- *Dynamic RAM* (DRAM).

- *Static RAM* (SRAM).

As mentioned, the difference between the two types of RAM is the technology used to store data. The SRAM unit uses components constructed of gates. This is very much like the circuits we have been building up to now. The DRAM unit uses *capacitors* to store data. A capacitor is an electrical component capable of storing a charge. It can be used to store a single bit. An empty capacitor would represent a 0 value, and a fully charged capacitor would represent a value of 1.

Capacitors are used in many circuits other than DRAM units. One which you are familiar with is the flash mechanism in a camera. When you take a picture with flash, you will notice that the flash unit is first charged. This unit contains a capacitor that stores up a charge, large enough to flash a very bright bulb. Charging the capacitor takes a certain amount of time. And, once the capacitor is charged, and if power is removed without activating the flash, it will immediately start losing its charge. It will eventually end up completely drained. The charge drainage process is called *leakage*.

Leakage of charge presents a problem when using capacitors to store bits; data stored with capacitors is gradually lost. To fix this, DRAM units must be *refreshed*. Refreshing a DRAM is a process in which all data in the DRAM, with bits that are starting to weaken, is read, and then rewritten, at full strength.

If we compare DRAM with SRAM, we might consider the following dimensions.

- Access Speed: DRAM units tend to be slower than SRAM. This is because charging capacitors requires a latency, whereas gate propagation times in SRAM are much smaller.[3]

- Density: Density is the number of bits that can fit in a physical space, or to say it another way, how compact the device can be made. It turns out that capacitors can be built much smaller than gates, and the DRAM can be built more compactly than the SRAM.[4]

- Cost: Cost refers to the cost per bit. Again the DRAM does well. Storage using capacitor technology is cheaper to build than storage using the technology used in gates.[5]

The comparison of DRAM to SRAM reveals something about the uses of these two types of RAM. SRAM is used in situations in which speed is at a premium. However, you can expect to pay a higher price. For situations in which a large amount of memory is required, DRAM is more appropriate, although you will need special circuitry to perform refreshing. SRAM is therefore used for specialized high-performance applications, and DRAM is used for normal working memory in the computer system.

4.1.6 ROM Types

Let us now look at ROM units. These can be classified as follows.

- ROM: Standard read-only memory.

- PROM: Programmable ROM.

- EPROM: Erasable PROM.

- EEPROM: Electrically EPROM.

ROM chips are read-only. To load the content into a ROM chip, essentially, the ROM chip is "printed" with the contents. Once printed, the chip is unalterable. The equipment to print ROM chips is a financial investment that

---

[3]SRAM units have access time as much as 6 times as fast as those in DRAM units.

[4]A DRAM unit can be up to 6 times denser than a SRAM unit with the same chip area.

[5]A SRAM unit can cost more than 50 times more than a DRAM unit of the same size.

only large-scale production houses would want to make. But then, if you have a smaller business that would like to produce a limited run of ROM chips, you would have to look for another alternative.

PROM chips are *programmable*. The meaning of programmable, however, is a bit different than what a computer scientist would think of. Programmable means that the contents of the ROM chip can be loaded after chip manufacture. To program a PROM chip, you would purchase a *PROM burner*. A burner is a small device into which a blank ROM chip is inserted. The device is connected to a host computer. At the host computer, the contents of the PROM are composed. When ready, the user downloads the program to the burner, which in turn burns the program into the PROM chip. Once the program is burnt into the PROM, the programming cannot be altered. If a modification is required, the only alternative that the user has is to discard the chip and burn a new one.

To solve the problem of non-reusable chips requires an EPROM. The EPROM chip is used, to a point, just as the PROM is used. That is, the EPROM is burnt by a PROM burner, connected to a host machine. However, if the user, at some point, decides to modify the programming, this is possible. Actually, the user cannot exactly modify the programming. Instead the user can wipe out the contents of the EPROM, transforming it into a blank chip again. The chip can then be reprogrammed with the modified contents.

An EPROM chip has a distinctive appearance. It looks very much like a PROM, except that on the top of the chip package is, typically, a clear plastic window, exposing the internal circuitry to view. To erase the chip, you would shine ultra-violet light through the window, for a certain amount of time. (The burner usually has a UV chamber for doing this.) The UV light releases any connections made by the programming, and returns the chip to its unused state. EPROMs, of course must be protected from stray, unwanted UV exposure, so often, when deployed, these chips will have a piece of opaque tape covering the window.

To reprogram an EPROM requires a UV source. The EEPROM chip can be reprogrammed in the field, without the use of UV light. In appearance, the EEPROM chip has no window. Instead it has a special pin used to erase the chip. To erase content on the chip, a relatively high voltage is applied to the erase pin. This high voltage resets the programmed connections, just as UV light does in the EPROM. The EEPROM technology is fairly flexible, and various versions of the chip exist. Different versions erase content in different units. That is to say, you might have a version that erases just a single word, a version that erases a block of words, or a version that erases the whole chip.[6]

4.1.7 Word and Byte Addressing

One problem with memory unit addressing has to do with data size. It is not uncommon for a processor to deal with several types of data of several sizes. As a motivating example, let's assume that a processor is doing arithmetic with integers. We would like to represent a large range of integers. We, as designers, have decided that a 16-bit word size is sufficient to represent all integers that we have interest in.

We also need to work with characters, like the characters that a user would type at a keyboard. We cannot directly work with characters, since all we are able to store is bits. So, we give each character a number, or as we might call it, a code. And, we discover that to represent all characters requires eight bits.[7] Eight bits is usually referred to as a *byte*.

We now have a problem. With two types of data, with sufficiently different word size requirements, how do we choose the word size for our memory unit? Do we choose a word size of 8 bits, and then how do we store the 16-bit integers? Or, do we choose a word size of 16 bits, and then how do we store a character code?

Suppose that we choose a 16-bit word size for the memory unit. We can easily store our integer values. Characters can also be stored in the word, although we would be wasting large amounts of space. For instance, suppose that we are storing a character with code 0110 0111. Stored in a 16-bit word this number would appear as 0000 0000 0110 0111. Notice that the high-order 8 bits are all 0. This is wasted space. In fact we could easily store two character codes in one word.

The other alternative was to choose a word size as one byte. But this means that we would not be able to store a 16-bit integer in one word. Our two extreme solutions both have problems. And so, we use a compromise solution. We use a 16-bit word. In the word we can store two characters or just one 16-bit integer. For example, if we wished to store two characters—first, 0110

---

[6]A flash unit is a type of EEPROM that erases content in blocks. Since flash is usually used to store files, and files are written and read in blocks, there is never a requirement to rewrite content smaller than a block, and so this system works well for file storage.

[7]Eight bits allows us to represent codes for 256 characters. Although there are not 256 characters on a standard keyboard, through the use of the modifier keys, like the shift and control keys, the number of character codes available on the keyboard is multiplied, so as to require eight bits.

0111, and next 0001 0101—we would store them in a single word as 0110 0111 0001 0101.

The solution we have proposed has solved some problems, but it has introduced a few more problems. If we examine how we might use such a memory unit, let's suppose that we are interested in writing the contents of a processor register, R0, into memory location 5. We might do this in assembly language with the following code.

```
mov M[5], R0
```

Now suppose that we wish to store the character code 36 in memory location 5. If we wish to store 36 in the lower byte of memory location 5, we can do it with the same assembly instruction. Bear in mind, however, that we will be setting the high byte to 0, which will cause problems if there is a character code already stored there. But the real problem occurs when we wish to store the value in the upper byte. We have no way of referring to the upper byte, since it does not have an address.

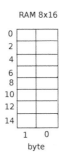

FIGURE 4.8 Memory with byte addessing.

The solution is to give every byte an address, not just every word. A 8×16 memory unit with byte addressing is shown in Figure 4.8. Each entry has two bytes. The word addresses, notice, are not incremented by one on consecutive lines, but rather by two. For example, the first word in the unit has address 0. This is also the address of the low-order byte of the word. The address of the high-order byte is 1. The second word in the unit has address 2. It cannot have an address of 1 because this is already used to address the upper byte in Word 0. We see that bytes are addressed sequentially, and word addresses are always the same as the byte address of their lower byte.

With this addressing scheme we now have solved our problem. To work with our addressing scheme, we would have two separate instructions for writing to memory: *movw*, for moving a full word, and *movb*, for moving a byte. To write a 16-bit value to memory location 6 would be done using the following assembly code.

```
movw M[6], R0
```

To write just a byte to the lower byte of location 6 would be

```
movb M[6], R0
```

and to write the upper byte would be

```
movb M[7], R0.
```

Notice that the assembly instruction

```
movw M[7], R0
```

is not allowed. Address 7 is not a word address, and we are performing a word move instruction. When performing a word move instruction we must specify addresses that are word addresses. In our example, word addresses must be multiples of two. This requirement on addressing is referred to as *alignment*. We say that addresses for word instructions must be aligned.

FIGURE 4.9 Byte addressing schemes.

4.1.8 Machine Byte Order

One last note on byte addressing has to do with the way the bytes are arranged in the word. Using an example to illustrate the concept, assume that we are working with a 32-bit word. This word would be split into 4 bytes, as shown in Figure 4.9. On the left of this figure we see the byte addresses of the individual bytes in Word 0, using the number scheme which we have used up to now. We start numbering bytes from the low-order byte, up to the high-order byte, with increasing addresses. This addressing scheme is called *little-endian* addressing. On the right is an alternative addressing scheme. Here, the high-order byte is given the low address, and byte addresses increase as we proceed to the low-order byte. This addressing scheme is called *big-endian*.[8] To remember which scheme is which, you think of where the low-address byte is located; in little-endian it is at the low-order, or little end of the word, and in big-endian it is at the high-order, or big end of the word.

Different computer models use different endian schemes for byte addressing. The choice, in many respects is arbitrary, but the difference often shows up when writing character string processing code.

4.2 PERIPHERAL DEVICES

The second type of external device connected to the CPU via the bus is the peripheral device. The range of peripheral devices is very large, and the devices can perform widely varying functions. To discuss these devices in detail would be a monumental task. So, instead, we try to simplify these devices, and consider only those characteristics concerned with interaction with the CPU.

4.2.1 Peripheral Device Types

In a simplistic way, we can categorize peripheral devices into one of three categories.

- Input devices. These are devices from which the processor reads data. The keyboard and pointer devices like the mouse are examples of input devices.

- Output devices. These are devices to which the processor writes data. The monitor and printer are examples of such devices.

- I/O devices. These are devices that combine both an input element and an output element. The processor can write to, and read from, these devices. An example of such a device is a disk drive.[9]

With this categorization we can now present the interface to the different categories of devices. Remember that we are showing only simplified versions of the actual interfaces.

Figure 4.10 shows the interfaces for the three device types. The top diagram is of an output device. The block labeled "Out" is the device *actuator*. It is the circuitry and machinery that process digital signals, to produce some form of

---

[8]The terms *little-endian*, and *big-endian* are taken from Jonathon Swift's *Gulliver's Travels*.

[9]It is quite appropriate to think of a memory unit as an I/O device. We have, however, made a distinction between devices that have internal addressing, calling them memory units, and devices that do not, calling them peripheral devices.

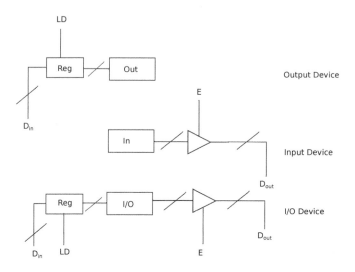

FIGURE 4.10 Peripheral device types.

output. The actuator receives input from a device register. Usually the output device requires some time to perform its task, and this register latches the data with which the output device is working, so that it is available to the device during processing. To use the device, the CPU places data on the D_{in} port of the device, and strobes the LD line, loading the register. The actuator then takes care of processing the data loaded. As an example, if the output device were a monitor, the CPU might place a character in the device register. The actuator of the monitor would then draw the character.

The middle diagram of Figure 4.10 shows the interface for the input device. The block labeled "In" is the device *sensor*, which collects input data to be sent to the CPU. The output of the sensor is connected to a switch. When the CPU requires data from the input device, it strobes the enable line on the switch, and reads the data from the D_{out} port of the device. As an example, when the user would type a character on a keyboard, that character would be sent out to the switch, which could be opened by the CPU.

The bottom drawing in Figure 4.10 is that of the interface to an I/O device. You can see that the I/O device is just a combination of both an input and an output device. The block marked "I/O" is a combination actuator and sensor, capable of consuming output or producing input. The device is read by activating the tri-state switch, and written to by placing data on the data-in port, and then strobing the load line.

4.2.2 Device Polling

For the most part, our simplified interfaces are workable. There are, however, some problems that we must discuss. A major problem has to do with getting the CPU to synchronize with the peripheral device. Let us consider a motivating example to introduce this problem.

It is not uncommon, when reading in characters from a keyboard, to read in sequences of characters, which are called *character strings*. The process is to read the characters, one by one, and store them in consecutive locations in memory, in a data structure called a buffer. We begin the process by triggering the E line on the keyboard input device. From the keyboard D_{out} port we read a character code. The code we read, let's say is a 65. This turns out to be the code for the character 'A'. we take this character and place it in our buffer. We then return to the keyboard and again trigger the enable line. This time we receive a 66 (the character 'B') as output. This is then placed in the next memory location in the buffer. Again we return to the keyboard, and strobe the enable line. This time we read out a 66, again. Now, there are at least two possible explanations as to why we received a 'B' twice.

1. The user has typed the character 'B', twice in a row.

2. The user has not had time to type another character yet, and we are still seeing the 'B' that was previously placed in the buffer.

It should be mentioned that the most likely of the two explanations is Explanation 2. This is because the computer is working so much faster that the user; a computer can execute, easily, millions of instructions per second, whereas a fast typist might, possibly, be able to type five characters per second. In any case, either of the two options is possible, and we need a mechanism to allow us to determine which explanation applies.

The mechanism used involves connecting each peripheral device to a *status register*. The bits in this register give information on the status of the device to the CPU. One of the bits in the register is often called the READY bit. This is the bit that allows us to solve the problem of determining the cause of a repeat character from the keyboard. On an input device, the READY bit is normally cleared. For the keyboard, when the user types a character, the character code is made available as output, and the keyboard controller raises the READY bit, indicating that input can now be read. When the CPU triggers the enable line, and reads the available character, the READY bit is cleared, automatically.

Let's go through our keyboard example again. The CPU, let's say, is ready to start reading in characters. However, before reading the first character, it checks the keyboard READY bit. If the bit is 0, then the CPU waits. That is to say, it enters a loop in which it repeatedly checks the READY bit, and only exits the loop when the READY bit raises to a 1. Then the first character is read, and placed in the buffer. The act of reading the character from the keyboard causes the READY bit to be cleared. The CPU then waits for the READY bit to be set again by the keyboard, indicating that the second character is ready to be read. The second character, a 'B', is now read in. It now attempts to read the third character. To read it, it waits until the READY bit is set. Notice that by waiting for the READY bit before reading, it knows, with certainty, that if then it receives a 'B', that 'B' was typed twice by the user. If it were the same 'B' as it read when reading the second character, the READY bit would still be a 0.

Output devices also have READY bits. The READY bit for an output device is normally a 1, indicating that the device is ready to receive data. When the CPU writes data to the device register, the READY bit is lowered to 0, and the device begins processing the data. When the output device has finished processing the data, the READY bit is automatically raised back to a 1. To send output to the device, before loading new data in the device register, the CPU would wait for the device READY bit to show a value of 1, to avoid overwriting data which the output device had not finished processing.

4.2.3 Interrupts

Although we have now solved the problem of determining if a device has finished processing new data, our solution is inefficient. The CPU now must wait for the READY bit to raise to a 1 value. As explained, this process of waiting is actually an active process, often referred to as a *busy wait*. That is to say that the CPU is, in fact, doing something while waiting. It is executing a loop, polling a device to see if it is ready. Since the processor is doing work during the polling, it cannot be used for anything else, as a processor can only execute one instruction at a time, in general.

Having established that the processor can do nothing else during a busy wait, the question rises as to what else it might be doing. Modern computers normally are working on several programs at once. This is evident in a window-based operating system. We might have several windows open at once, each window doing something different. Each of those windows is controlled by a separate process, or program. The CPU is executing all of these processes concurrently.

Stating that several processes are executing simultaneously is a little misleading. Remember that a processor can only execute one instruction at a time. However, to the user, it appears that the processor is executing several programs concurrently. To present this illusion, what actually happens is that the processor jumps from one process to another, executing a little piece of each process, before moving on to another process. If the amount of time that the processor works on each process, often called the *quantum*, is small enough, the user does not notice that a particular process is suspended, because the processor returns to the suspended process before the user notices the suspension. This method of simulating concurrency, by switching from process to process, is often called *time division multiplexing.*

Notice that if the processor is locked up in a busy wait, time division multiplexing is not possible. So, we start looking for a mechanism that can eliminate the busy wait polling loop. The mechanism typically used is based on the concept of an *interrupt*. To understand interrupts, let's look at an analogy. Imaging that you are trying to weave a basket. You are also expecting a delivery at your door. You are having trouble concentrating on the basket weaving, because every two minutes you have to look up to see if the delivery person is at the door. What do you do? Well, most likely what you do is install a door bell. Now you do not need to keep glancing at the door. Instead, when the delivery person arrives, they ring the door bell, and you are sent an auditory signal, notifying you of the event. You can then suspend your basket weaving activity, go to the door, receive the delivery, and then return to the basket weaving.

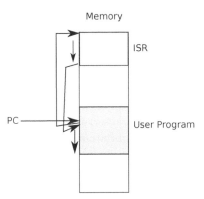

FIGURE 4.11 Interrupt mechanism.

Let's apply this type of mechanism to a processor. Suppose that the processor is executing a program. Simultaneously, it is collecting characters from a keyboard. When a character is ready, it needs to be copied into a buffer. When a character comes in, the program executing must be suspended, another program that copies the character from the device to the buffer must be executed, and then the processor must return to the suspended program and continue executing it, exactly where it left off. This suspension process, called an interrupt, starts up when the processor receives an interrupt signal from the keyboard device. The interrupt process is illustrated in Figure 4.11.

Figure 4.11 shows memory. The CPU is currently working on a program. Remember that to execute, a program must be loaded into memory. The workspace for the program is shown, and marked "User Program." The figure shows the location of the PC. So, at this time the CPU has executed a small portion of the program, and the PC is now pointing at the next instruction to execute. At this point in time, however, an interrupt signal is received by the processor. It then suspends execution of the user program, and jumps down to another program, labeled "ISR" on the diagram, and begins executing this program. The Interrupt Service Routine (ISR), often called the *interrupt handler*, is a piece of code which is part of the operating system. In this sense, you can think of it as part of the computer system. For a keyboard interrupt, the interrupt handler would execute code to fetch the character code from the keyboard device, and copy it into the keyboard buffer. After finishing this, it would jump back to the user program, and continue from where it left off. The interrupted program is unaware that it has been interrupted at all. Any registers in use by the program are left unchanged, or if they are needed by the ISR, they are saved when the interrupt occurs, and restored when the ISR returns control to the user program.

There are many causes of interrupts. In our example, we assume that the keyboard causes the interrupt. However, any device on the bus can cause an interrupt. This type of interrupt is called an *external interrupt*. Further, even parts of the CPU might send an interrupt signal to the CPU. These interrupts are called *internal interrupts*. With all of the different causes of interrupts, the ISR has to determine the cause of an interrupt before it can execute the correct piece of code, to perform the appropriate action. The way this is handled, is that the device issuing the interrupt signal would not only raise a signal line, informing the CPU of the interrupt, but even before this it would place a code in a CAUSE register. The code is a number which indicates the type of interrupt. So, for instance, a code of 8 might indicate a keyboard interrupt, and a code of 15 might be a mouse interrupt. When the ISR starts up, one of the first things it does is check the CAUSE register, to determine what to do.

Interrupts are often requested by devices either internal to the processor, or external to the processor. It is also possible for the user's program to request an interrupt. These types of interrupts are called *software interrupts*. You might wonder what would cause a program to ask the operating system for an interrupt. The answer is that most programs run under a restricted *user mode*. In user mode, a program would not be able to directly manipulate many peripheral devices to, let's say, send output to a printer. To do this, a process must be executing in *kernel mode*. In kernel mode there are no restrictions on what the process can do. The operating system runs in kernel mode. So, to print something, the user program must ask the operating system to do this in its stead. This is done by asking for the operating system to interrupt it, and passing it information telling the operating system that it wants some information printed. The ISR then takes over, prints the given output, and returns to the user program.

4.3   THE CPU

The last device connected to the bus is the processor. We often say that the processor is a device that executes programs. It might be more accurate to say that the CPU is a device that executes machine instructions. As long as the power is on, a processor fetches instructions and executes them. Those instructions often cause the CPU to write and read data, to and from memory and the other peripheral devices on the bus connection.

We just made the statement that as long as the power is on, the CPU fetches and executes instructions. We can rephrase that with a little more accuracy, by saying that as long as the power is on, the CPU executes a cycle, called the *machine cycle*, or *instruction cycle*. In this cycle, it fetches an instruction, figures out what action the instruction is asking for, and performs the requested action. The processor then repeats this cycle, over and over again. In this sense, the processor is a rather simple machine. It does not do anything too involved; it simple performs the same cycle, mechanically, and repeatedly.

When we write programs, they are translated eventually into machine language. The machine code is stored in a file on disk. When the program is run, the code is loaded into a workspace in memory. This is all performed by different parts of the operating system. The program is then executed from memory.

The machine cycle, or instruction cycle, is a sequence of steps used to execute one machine language instruction. These steps are the following.

1. *Fetch.* The PC contains the address of the next instruction to be executed. The instruction indicated by the PC is fetched from memory. The instruction is brought into the CPU, and placed in a register. The PC is also updated to point to the new next instruction.

2. *Decode.* The instruction is examined. The processor determines the operation to be performed, and the location of the operands required.

3. *Execute.* Any operands in memory, registers, or coming from a peripheral device are fetched, and the operation is performed. If there is a result produced by the operation, it is stored in the specified destination.

The three steps of the machine cycle are performed repeatedly, to execute a whole program.

To illustrate the workings of the machine cycle, consider the following example. Suppose that the PC currently contains the address 4000. Let us follow through one iteration of the machine cycle. During the fetch phase of the cycle, the instruction at memory location 4000, M[4000], is fetched into the processor. Suppose that that instruction fetched, in binary, is

    10000, 0000, 00011110

or in assembly language

    add R0, M[30].

After reading the instruction, the PC is updated to point to the next location in memory, 4001. Next we enter the decode phase. The processor circuitry realizes that the op-code 16, from the first field of the instruction specifies an *add* instruction. Then it determines that one of the operands is in R0, by examining the second field. The other operand, given by the third field, will be fetched from M[30]. In the execute phase, the operand in memory is fetched into the processor, it is added to R0, and the result is placed in R0, completing one iteration of the machine cycle.

4.4 BUS COMMUNICATION

We see that, in the machine cycle, values are constantly being fetched from memory or peripheral devices. The results of an operation are also often written out to bus devices. In this section, we discuss the circuitry involved in read, and write operations over the bus.

4.4.1 Bus Structure

Referring back to Figure 4.1, we have drawn the bus as a single cable. That single cable is shown again in Figure 4.12. The figure magnifies the bus cable, showing that it is composed of three sub-buses: the *address bus*, the *data bus*, and the *control bus*, labeled A, D, and Ct, respectively.

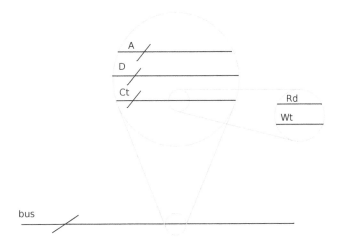

FIGURE 4.12 The bus and sub-buses.

The address bus is used, by the CPU, to send addresses to devices on the bus. Memory is one possible device on the bus, and when using a memory unit, we need to send it an address. The address bus is used to do this. But its use is even more sophisticated than this. We can view all of the devices on the bus as comprising a giant virtual memory unit. That is to say, if there is a monitor as one of the devices connected to the bus, it is considered as part of this virtual memory unit, and has an address. If a keyboard is connected to the bus, we would also consider it as part of this virtual memory unit, and assign it an address. These virtual addresses, or more correctly *bus addresses*, are what is sent over the address bus.

The data bus is used to pass data to, and receive data from devices connected on the bus. The type of data transferred over the data bus is device dependent. So, for example, if the CPU is performing a write to memory, it would be sending data to be stored at the specified location, and if it were reading from a mouse it might be receiving mouse coordinates.

The control bus is used, by the CPU, to send control signals to, and receive status signals from the devices on the bus. For example, a RAM unit would have control signals W and E. An input device might send a READY bit to the CPU. All of these signals would be sent over the control bus. In general, control signals are sent out to devices to tell them what operation to perform, and signals are received from devices indicating the status of an operation.

We use a very simple control bus in our discussion, to illustrate bus addressing. Figure 4.12 shows that our control bus is split further into two lines. In general, the control bus may carry many more than two lines, but we have simplified the control bus to just two lines: the read line, Rd, and the write line, Wt. When the CPU asks a device for a read operation, the Rd line is raised. To request a write operation the Wt is raised.

4.4.2 Bus Addressing

Let us examine what is required of the CPU to request a write operation from a memory unit. First, the data being written must be placed on the D bus, and the address to which it will be written must be placed on the A bus. Then the CPU strobes the Wt line. The memory unit sees the Wt line raise, and becomes aware that it should be doing a write operation. But, then we see a problem. There may be several memory units on the bus. They all see the request for a write operation, and they all will perform a write. At the core of the problem is that read and write messages, from the CPU, are being broadcast. All bus devices see the message. We need some way for each device to determine if the message is meant for itself, or some other device. Messages for itself would be processed, and messages for other devices would be ignored.

The method we use for identifying the recipient of a message has to do with the bus address. The bus address, as mentioned, is an address in a larger virtual memory that encompasses all bus devices. In order to simplify our address decoding circuitry, we spit our bus address into two fields: the unit number, and the internal address.[10] If this sounds familiar, it's exactly the same method used in Figure 4.6, to implement composition of several memory units. In fact to build our virtual memory, we are composing general devices, rather than just memory units.

We assign a unit number to every bus device. When the CPU sends a message over the bus, it sends a full bus address, including an internal address, which would be used only by memory units, not peripheral devices, and a unit number. Each device checks the unit number that was broadcast against its own unit number. If the unit numbers match, then the received message is processed, and if not the message is ignored.

---

[10]Our address decoding scheme is an example of *partial address decoding*, in which not all of the virtual address space is implemented. In *full address decoding*, all of the address space is implemented. Although partial decoding can produce simpler circuitry than full decoding, full decoding makes more efficient use of the address space.

4.4.3 Bus Addressing Example

Now, we must implement this matching, or *address decoding* strategy in circuitry. We illustrate how this is done with an example. Consider the following system specification.

- An 8 × 4 RAM unit; addresses range from 0000000 to 0000111.

- A 16 × 4 ROM unit; addresses range from 0010000 to 0011111.

- An input device with address 0100000.

- An output device with address 0110000.

- An I/O device with address 1000000.

We can learn a lot of information about this machine from the specification. First, we learn that the word size on this machine is 4 bits. We can usually deduce this from the width of the memory units. This, in turn, tells us that the data bus, D, will be 4 bits wide. We also see that there are five devices on the bus. If we number these devices 0 through 4, we will need a 3-bit unit number (000 through 100). The RAM unit has a length of 8, and so requires an address of 3 bits. The ROM has a length of 16, and so requires an address of 4 bits. The internal address field of the bus address must be large enough to accommodate both the address for the RAM, and the addresses for the ROM. So, we make the internal address 4 bits, to accommodate the larger of the two addresses, that of the ROM unit.

We now know quite a bit about the bus address. The unit number will be 3 bits. The internal address will be 4 bits. The total size of the bus address will be $3 + 4 = 7$ bits. You can see that the addresses given in the specification for the devices are, in fact, all 7 bits. If we examine these bus addresses closely now, they start to make sense. For instance, the bus addresses for the RAM are 000 0000 through 000 0111. Using the three high-order bits as the unit number, as we do for high-order interleaving, we discover that the unit number for the RAM is 000. The RAM is Unit 0. We see that the internal addresses range from 0000 to 0111, or in decimal 0 through 7, as we would expect for a unit with a length of 8. Analyzing the ROM unit in the same way, we see that its unit number is 001, indicating that it is Unit 1, and its addresses range from 0000 to 1111 (decimal 15), as we would expect for a unit of length 16.

Moving on to the peripheral devices, the input device has a bus address of 010 0000. This would be Unit 2. The internal address is irrelevant, since the device has no internal addressing. The output device, with bus address 011 0000, is Unit 3, and again the internal address is not used, and finally, the I/O device has bus address 100 0000, indicating it is Unit 4, and again it has no internal addressing.

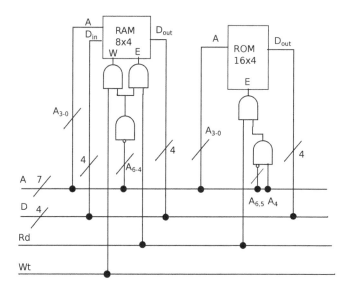

FIGURE 4.13 Memory connection to the bus.

We are now ready to start connecting the devices to the bus. All five devices are connected to the bus. In Figure 4.13 we show how the memory units are connected. The address bus is the same width as the size of the bus address. The data bus is the size of the computer word. The control bus is shown split into the read and the write lines.

The ROM unit has its ports connected to the appropriate buses. The data-out port is connected to the data bus. When the read operation is performed, the data produced by the ROM will flow onto the data bus, allowing the CPU to read it. The address port is connected to the address bus, but only the bits that constitute the internal address, Bits 0 through 3.

The E control line of the ROM, in Figure 4.13, is wired to a control circuit that checks for two conditions: it checks that the CPU is requesting a read operation, and that the unit number being sent by the CPU is 001. The control circuit is composed of two gates. The lower gate is connected to the unit number field coming off the address bus. These are the lines A_6, A_5, and A_4. A_6 and A_5 are sent through inverters to check if they are 0. A_4 goes into the AND gate as is, checking that it is a 1. So, the lower gate, which we call the *addressing gate*, checks that the unit number portion of the bus address is 001, specifying Unit 1. If it detects the proper unit number, it outputs a 1 into the top AND gate, called the *operation gate*. This AND gate will fire if the addressing gate fires, and the Rd line has a 1 on it, specifying a request for a read operation.

This structure, with an addressing gate feeding into an operation gate, is used for all devices on the bus. If we look at the RAM unit of Figure 4.13. we see the same structure. The addressing gate is connected to the unit number coming off the address bus. All three bits are sent through inverters. This gate, then, fires only if the unit number is a 000, indicating Unit 0. There are two operation gates: one connected to the W line, and one connected to the E line. The operation gate for the W line is connected to the addressing gate, as always, and the Wt line on the control bus. It fires only if the request is for Unit 0, and the operation requested is a write. The operation gate for the E line is connected to the Rd line, checking for a read operation request for Unit 0.

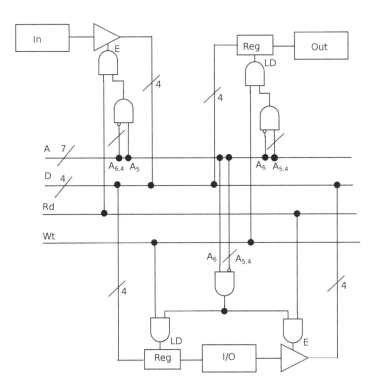

FIGURE 4.14 Peripheral device connection to the bus.

The other lines of the RAM are connected in a standard fashion also. The address port is connected to the internal address portion of the bus address, and the data lines are connected to the data bus. This pattern of connection is followed for any device connected to the bus.

Figure 4.14 shows the peripheral devices, for our example, and how they are connected to the bus. The input device has its data-out port connected to the data bus. Its addressing gate checks for unit number 010 (Unit 2), and its operation gate checks for a read operation. The output device has its data-in port connected to the data bus. Its control circuitry checks for a unit number of 011 (Unit 3), and a write operation request. The I/O unit has both data-in and data-out connected to the data bus, and its addressing gate checks for unit number 100 (Unit 4). The operation gate for its load line checks for a write operation request, and the operation gate for its enable line checks for a read request.

Notice that, as previously discussed, peripheral devices do not use the internal address portion of the bus address. This is an interesting point. We are told that the bus address for the input unit is 010 0000. And so, from that perspective it would make sense for the addressing gate to check both that the unit number is 010, and that the internal address is 0000. However, checking the internal address creates control circuitry that is not needed, since the unit number, in our simple scheme, uniquely determines the device.

If you have many devices, you may need to give several devices the same unit number. In that case you can use the internal address as a *sub-unit number*. For example, you might have two devices with unit number 111. You might then give one device the sub-unit number 0000, resulting in a bus address of 111 0000, and the other device the sub-unit number 0001, resulting in a bus address of 111 0001. In this case the addressing gate would need to check both the unit number and the internal address being used as the sub-unit number.

4.5 SUMMARY

In this chapter, we have presented a view of the bus architecture found on many computers. This architecture is not the only architecture possible, and there are several other architectures in use. Other architectures might be classified as direct connection architectures, in which each device is directly connected to a set of other devices. The direct connection architecture has the advantage of speed. The increased speed is a result of a higher degree of concurrency.

The bus architecture consists of a single connection between devices. Messages are sent using a broadcasting scheme. Devices listen to the bus for messages for which they are the intended recipient. These messages are processed, and other messages are ignored.

We discussed the types of devices on the bus. Besides the CPU, these devices can be classified as memory devices with addressing, and peripheral devices involved with non-addressed I/O.

A memory device can be thought of as an array of words. Each word stores a single multi-bit number, and is identified by its address. Memory units were classified, by operation, into the classes ROM and RAM. RAM units are further classified by storage technology, with the classes SRAM and DRAM. Further classification of ROM units is by programmability. This yielded the classes PROM, EPROM, and EEPROM.

We showed the structure of a memory unit, and discussed how we might compose units into larger units. The discussion of memory units ended with a section on word and byte addressing.

Our discussion then turned to peripheral devices, without addressing. We presented a simple, standard interface for these devices. We then discussed how these I/O devices are used. We talked about using active status bit polling to determine if a device is ready for further I/O. It was observed that this technique is inefficient, and so we then discussed using interrupts to handle I/O. This scheme, rather than being purely software based, has hardware assistance. In the interrupt scheme, rather than the CPU actively polling the device, the device notifies the CPU when it needs service. When the CPU receives the interrupt notification, it suspends its current work, handles data transfers for the device, and then resumes its suspended work.

The last bus device discussed was the CPU. We did not cover much on the processor; we do not yet have enough background to fully cover it. What we did cover was the machine cycle. This sequence of steps, in a nutshell, is what the processor does.

The chapter ends with a discussion of bus connection. We talked about the three sub-buses that comprise the bus: the data bus, the address bus, and the control bus. We discussed how each device is assigned a bus address, that uniquely identifies the device. The bus address is sent out by the CPU with each message it sends. We then showed how the control lines for a device might be wired to respond only to messages with the device's proper bus address.

With this discussion, we now have a better understanding of the environment in which the CPU functions. In fact, with knowledge of the machine cycle, we are almost ready to design a simple processor.

4.6   EXERCISES

4.1   Given an ample supply of 4 × 4 ROM chips, compose them into the following configurations.

    a.   Draw a 16 × 4 ROM unit.

    b.   Draw a 4 × 16 ROM unit.

    c.   Draw an 8 × 8 ROM unit.

4.2   The types of RAMs not discussed in this chapter include the SSRAM and the SDRAM. Investigate these memory types. Write a description of them, and give a simplified block diagram showing their interfaces.

4.3   You are working with a 256 × 32 RAM unit. For the word M[64], you are trying to shift the bytes to the right, performing what is called a *right circular shift*. In a right circular shift, each byte would shift one byte to the right, and the byte at the very right of the word would wrap around and replace the high-order byte. As an example, if M[64] contained 00000000 10101010 01010101 11111111, after the circular shift right it would contain 11111111 00000000 10101010 01010101. Write assembly code to implement the circular shift right, using the *movb* instruction discussed in this chapter.

    a.   Assume that the machine you are working on is little-endian.

    b.   Assume that the machine you are working on is big-endian.

4.4   An interesting question is, what happens if the CPU is interrupted while executing the ISR? Think about what this means; the CPU receives a new interrupt before it is through processing another.

    a.   Why might a you not want this to be allowed?

    b.   Why might you want this to be allowed?

    c.   Based on your answers to Parts 4.4a. and 4.4b., if you gave the user the ability to disable interrupts, would you want to give the user the ability to disable all interrupts, selected interrupt types, or both all and some interrupt types?

4.5   Given a standard bus connection with two control lines, *Wt* and *Rd*, show the wiring diagram for the following specifications.

- A 32 × 4 RAM unit, with addresses 0100000 to 0111111.

- An output device, with address 1000000.

- An input device, with address 1100000.

4.6 Given a standard bus connection with two control lines, Wt and Rd, show the wiring diagram for the following specification.

- An I/O device, with address 000000.

- A 16 × 8 ROM, with addresses 010000 to 011111.

- An I/O device, with address 110000.

4.7 Given a standard bus connection with two control lines, Wt and Rd, and a 4-bit address bus, show the wiring diagram for the following specification.

- An 8 × 4 ROM unit, with addresses 0000 to 0111.

- An input device. You should choose the bus address.

- An output device. You should choose the bus address.

- An I/O device. You should choose the address.

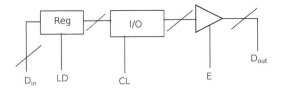

FIGURE 4.15 A resettable I/O device.

4.8 A more complex version of the I/O device is the *resettable I/O device*. This device accepts three types of requests: a read request, a write request, and a *reset* request. The diagram of the device structure is shown in Figure 4.15. It has three control lines: LD and E, as would any I/O device, and CL, which resets the device. The bus being used only has the two control lines Rd and Wt. To request a reset, the CPU simultaneously sends both the read and write requests. For a 4-bit data bus, an 8-bit address bus, and 3-bit unit numbers, draw the bus connection diagram for this device, if its unit number is 110.

CHAPTER 5

The Register Transfer Language Level

CONTENTS

W<small>E ARE NOW AT THE POINT</small> where we can describe low-level circuitry in the processor. The problem is that we need to start describing higher-level circuitry in the processor, and the tools that we have developed for circuit description are cumbersome at the higher level. What makes our tools less adept at describing more complex circuitry is that these tools all provide a structural description of the circuit. That is to say, they describe the circuit in terms of what it is composed of, and how the pieces are connected. While this type of description provides a good precise specification of a circuit, when dealing with complex circuitry, we need a description that is more abstract, so that we can ignore some of the detail during the design process.

A more abstract description of a circuit is provided by a behavioral description. With a behavioral description, rather than describing how a circuit is composed, we describe what operations the circuit performs. In this chapter we introduce a common description language, called *Register Transfer Language* (RTL). This language describes the operation of a circuit using *micro-instructions*. A collection of these micro-instructions is often called a *micro-program*.

The terms *micro-instruction* and *micro-program* are a little misleading to someone with a background in computer science. The tendency for a computer scientist is often to think of micro-instructions as, somehow, representing executable code. Later we will see that this way of thinking actually has some validity. But, for now, it is much more important to think of a micro-instruction as a piece of circuitry. This means that each RTL micro-instruction can be translated into a circuit schematic. Likewise, it is possible to translate any circuit diagram into an RTL description. This is simply saying that any circuit, either high level or low level, can be specified behaviorally or structurally.

5.1 MICRO-INSTRUCTIONS AS CIRCUITS

The behavioral description language we are discussing is called Register Transfer Language. The name implies that the language describes how data is moved from one register to another. Although RTL can show more than register-to-register movement, this is actually a fairly good characterization of RTL. As it turns out, most of the work that a processor does is moving data from register to register. This is the case because a processor, typically, is wired so that to perform a particular operation, the data must be located in special registers. This is done to simplify the processor design, but the result is that, to perform a sequence of operations, data must be moved from the result register for one operation into the operand registers of the next operation.

5.1.1 RTL Design

RTL descriptions are composed of micro-instructions. As an example, consider the following micro-instruction.

$$T : R1 \leftarrow R2 \tag{5.1}$$

A micro-instruction is composed of two parts: circuit control and data-path specification. These two sections are separated by a colon. In the micro instruction of Example 5.1, the circuit is controlled by the input T. The data-path specification indicates that the contents of a register, R2, should be copied to the register R1. The behavior described by the micro-instruction can be summarized as follows.

If T is set, then move R2 to R1.

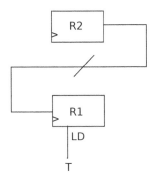

FIGURE 5.1 Circuit diagram for Example 5.1.

If we were to draw a schematic of the circuit specified in Example 5.1, the result would be Figure 5.1. Figure 5.1 is drawn in two steps. Usually we start by drawing the data-path. The data-path is composed of the connections, and components necessary to perform the operation specified by the micro-instruction. In this case, we draw two registers, and a connection from the output of R2 to the input of R1.

Once the data-path has been drawn, we then draw the control circuitry. To perform the specified operation, the register R1 must be opened to receive the data coming from R2. This is done by triggering the load line for R1. This should only be done when T is set. As a result, we have connected the T line to the load line of R1, with the result that if $T = 1$, R1 will receive the contents of R2.

Let us examine another example.

$$ab + \bar{c} : R1 \leftarrow 0 \tag{5.2}$$

The interesting feature of Example 5.2 is that control is a bit more complex. The example illustrates the fact that control can be any Boolean function of

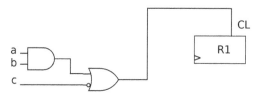

FIGURE 5.2 Circuit diagram for Example 5.2.

the control inputs. The structure of the circuit is shown in Figure 5.2. Notice how the control circuit uses gates to compute the value of the expression $ab + \bar{c}$. Also notice how we use the register control line, CL, to clear the register without external circuitry.

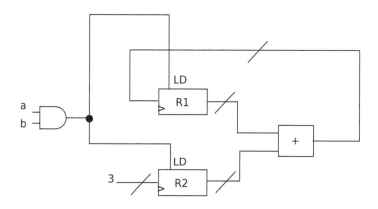

FIGURE 5.3 Circuit diagram for Example 5.3

Up to now we have seen micro-instructions with only one *micro-operation*. That is to say, all micro-instructions we have seen so far have only performed one transfer. In the following example we have a micro-instruction with two micro-operations.

$$ab : R1 \leftarrow R1 + R2, R2 \leftarrow 3 \tag{5.3}$$

The micro-operations are separated by a comma. This micro-instruction indicates that, simultaneously, the sum of R1 and R2 should be loaded into R1, and R2 should be set to 3. The schematic for the circuit is given in Figure 5.3. You will notice the use of the adder to compute R1 + R2.

It is important to be a little more precise about what we mean when we say that two micro-operations are performed simultaneously. If you examine the diagrams in this chapter, you will see that each of the registers is connected to the clock. As a result, each operation specified by the control circuitry will be performed at the same clock trigger edge. This is our definition of simultaneity. That is to say, all micro-operations in a micro-instruction are performed in one clock cycle, and all at the same trigger edge.

Let's now look at another example.

$$ab : R1 \leftarrow R1 + R2$$
$$a\bar{b} : R1 \leftarrow 3 \qquad (5.4)$$

Example 5.4 differs from previous examples in that a register, R1, is being loaded with a choice of values. In one case it is loaded with the sum of R1 and R2, and in the other case it is loaded with the constant 3. The choice is determined by the values of the variables a and b.

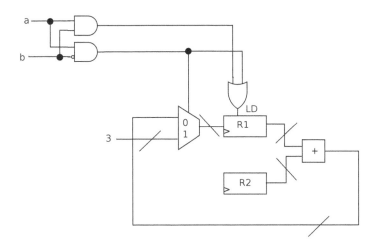

FIGURE 5.4 Circuit diagram for Example 5.4.

The common way of building choice into a circuit is with a MUX. The structure of such a circuit is shown for Example 5.4 in Figure 5.4. In this diagram, the data-path contains an adder to calculate R1 + R2. The value produced by the adder, and the value 3 are fed into R1, through a MUX. The MUX chooses between the two, and is operated by the trigger gates connected to a and b. In addition, the load line on R1 is controlled by an OR gate. This is because R1 must be loaded for the case ab, and also for the case $a\bar{b}$.

We can walk through the operation of the circuit in Figure 5.4 to illustrate how it works. When ab is true, the load line of R1 is switched on. Because the MUX selector line is not set, the MUX will allow Option 0, the value $R1 + R2$, to enter R1, performing the micro-operation $R1 \leftarrow R1 + R2$. When $a\bar{b}$ is true, the load line on R1 is again triggered, but so is the MUX selector line. This causes the MUX to allow Option 1, the value 3, to pass through, and this will be the value entering into the unlocked R1 register.

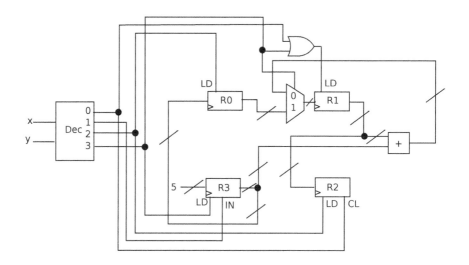

FIGURE 5.5 Circuit diagram for Example 5.5.

5.1.2 A Larger Example

We now present a larger example. This example puts all of our RTL circuit-building discussion together.

$$\begin{aligned}
\bar{x} \cdot \bar{y} &: R1 \leftarrow R1 + R3, R2 \leftarrow 0 \\
\bar{x}y &: R3 \leftarrow R3 + 1 \\
x\bar{y} &: R2 \leftarrow R1, R0 \leftarrow R3 \\
xy &: R1 \leftarrow R0, R3 \leftarrow 5
\end{aligned}$$
$$(5.5)$$

The circuit diagram for this larger example is shown in Figure 5.5. The diagram was constructed in the same way we constructed the diagrams for the smaller RTL examples. We started by drawing the data path. The circuit has four registers. R1 must be connected to an adder, together with R3, and the result of the adder must be wired into the input of R1. There must be a path from R1 to R2, and from R3 to R0. The constant 5 must be wired into register R3, and there must be a path from R0 to R1. Because R1 can receive inputs from both the adder and R0, we use the MUX and OR gate construct to deal with this, much as we did in the Example 5.4.

After we have laid out the data path, we then start work on the control circuitry. In previous examples we used trigger gates to trigger the proper micro-operation. A more compact way of triggering employs a decoder. The control signals, x and y are fed into the decoder and interpreted as a binary number, with x as the high-order bit, and y as the low-order bit. For example, if $x\overline{y}$ is true, that would imply that $x = 1$ and $y = 0$, which would be interpreted by the decoder as the binary number 10 (decimal 2). The decoder would then trigger its Option 2.

A thorough examination of the control circuitry in Figure 5.5 reveals that Option 0 triggers the load line on R1, and the clear line on R2. This implements the first micro-instruction for $\overline{x} \cdot \overline{y}$. Option 1 triggers the increment line on R3, implementing the micro-instruction $\overline{x}y : R2 \leftarrow R2 + 1$. Option 2 triggers the load lines of R2 and R0, implementing the micro-instruction for $x\overline{y}$, and finally, Option 3 triggers the load line on R1, the selector line for the MUX, and the load line on R3, implementing the micro-instruction for xy.

5.1.3   RTL Analysis

In this chapter, we have explained how to build circuitry from an RTL description. This process is the design process. The opposite process, the analysis process, is to build an RTL description, given a circuit diagram. To illustrate the analysis process, suppose that we are given the circuit diagram in Figure 5.5. To determine the RTL instructions we start at the decoder. We follow each trigger line out of the decoder, and record all micro-operations performed by that trigger line. As an example we will trace the Option 3 line.

As we follow the Option 3 line from the decoder, we see that it branches. The top branch triggers the MUX and the load line on R1. This means that R1 will receive the Option 1 input of the MUX, implementing the micro-operation $R1 \leftarrow R0$, since the MUX Option 1 line is connected to R0. The bottom branch of the trigger line triggers the load line on R3. Since the R3 input is connected to a constant 5, this implements the micro-operation $R3 \leftarrow 5$. The full data path part of the micro-instruction will be $R1 \leftarrow R0, R3 \leftarrow 5$.

Once we know the operations that the micro-instruction performs, we can then fill in the control signals. For Option 3 of the decoder, this line is triggered when the input is the binary number 11 (decimal 3). This is true when the Boolean expression xy is true. The full micro instruction is then

$$xy : R1 \leftarrow R0, R3 \leftarrow 5$$

as we would expect.

The above process would now be repeated for each of the remaining three trigger lines emanating from the decoder in Figure 5.5. When finished, we would have the four micro-instructions of Example 5.5.

5.1.4   Transforming a Structural Description into a Behavioral Description

Our procedure for transforming an RTL description into a circuit schematic is universal. Given any RTL description, we can use our procedure to derive a structural description. The procedure for converting from a schematic to an RTL description, however, only worked because the schematic of Figure 5.5 was drawn from an RTL description. As such, the circuit had a specific structure; it could readily be decomposed into data path and control components. In general, this is not the case, and so, as of now, we have no good way of converting from a schematic diagram to an RTL description.

Fortunately, there is a more general method for analysis. However, this more general method is complex to perform manually, even on some rather simple circuits, like our previous examples. As such, it is best left to circuit authoring software. But we can, and do, illustrate how it works with a simple example.

From Section 3.2.4, we know how to analyze a circuit schematic. If the circuit is a combinational circuit, we copy out Boolean equations for every output, and then build a truth-table using the equations. If the circuit is a sequential circuit, you would copy out equations for all flip-flop inputs and all circuit outputs, and then build the transition table from these equations. This method of analysis works well for the low-level circuitry we were dealing with in Chapter 3, but it can also be applied to higher-level circuitry. The problem is that, as stated, with high-level circuitry, the complexity of the task becomes overwhelming, and is better left to automation, rather than performed manually.

Our reason for discussing the analysis method of Section 3.2.4 is that we can assume that when we are producing RTL from a circuit schematic, we have already analyzed the circuit, and have developed a table specification of its operation. That table might be a truth table, or a transition table. In our example we will consider a transition table, since sequential circuits are more general than combinational circuits. The transition table we will use is that shown in Table 5.1. You might recognize this as the example FSA0 from Chapter 3.

TABLE 5.1 State transition table for example FSA0.

i	$Q_{(0)1}$	$Q_{(0)0}$	$Q_{(1)1}$	$Q_{(1)0}$	p
0	0	0	0	1	1
0	0	1	0	1	1
0	1	0	0	1	1
0	1	1	0	0	0
1	0	0	0	0	1
1	0	1	1	0	1
1	1	0	1	1	1
1	1	1	1	1	0

Remember how sequential circuits were implemented in Chapter 3: as a state register, which we will call Q, and a combinational circuit that calculated the new state for the register, and also the value of the circuit outputs, given the current state, and the circuit inputs. In example FSA0, the combinational circuit, would calculate $Q_{(1)}$ and p from $Q_{(0)}$ and i.

We will develop our RTL description using the transition table as our guide. Each row of the table will result in a micro-instruction that defines the output, p, and the new state, $Q_{(1)}$.

The first line of the transition table in Table 5.1 specifies that if i is 0, and the current state $Q_{(0)}$ is 0, then the new state, $Q_{(1)}$, should be 1, and the output, p, should be 1. In RTL this would be written as

$$\bar{i} \cdot \overline{Q_1} \cdot \overline{Q_0} : Q \leftarrow 0, p \leftarrow 0.$$

In this notation, the values Q_1 and Q_0 specify the high-order and low-order bits currently stored in the Q state register. We can continue producing RTL in this fashion, resulting in the following RTL specification.

$$
\begin{aligned}
&\bar{i} \cdot \overline{Q_1} \cdot \overline{Q_0} : Q \leftarrow 1, p \leftarrow 1 \\
&\bar{i} \cdot \overline{Q_1} Q_0 : Q \leftarrow 1, p \leftarrow 1 \\
&\bar{i} Q_1 \overline{Q_0} : Q \leftarrow 1, p \leftarrow 1 \\
&\bar{i} Q_1 Q_0 : Q \leftarrow 0, p \leftarrow 0 \\
&i \overline{Q_1} \cdot \overline{Q_0} : Q \leftarrow 0, p \leftarrow 1 \\
&i \overline{Q_1} Q_0 : Q \leftarrow 2, p \leftarrow 1 \\
&i Q_1 \overline{Q_0} : Q \leftarrow 3, p \leftarrow 1 \\
&i Q_1 Q_0 : Q \leftarrow 3, p \leftarrow 0
\end{aligned}
\tag{5.6}
$$

In summary, to convert from a circuit schematic into RTL, you would first analyze the circuit, producing a truth table, or transition table. Then, from the table you can read out the RTL instructions which produce circuit output and flip-flop input.

5.1.5 Problems with Reverse Engineering

The micro-program from Example 5.6 is what we require. We were interested in producing RTL from a schematic diagram, and this is what we have done. It is, however, still instructive to do a little more with this example. What we will do now is take our micro-program and build a circuit diagram from it. This is, of course, not necessary, since presumably, we already have a circuit diagram for the circuit; this is what we started with in Figure 3.31. However, producing a circuit diagram using our usual technique, from the RTL code of Example 5.6, reveals something interesting about the relationship between behavioral descriptions and structural descriptions.

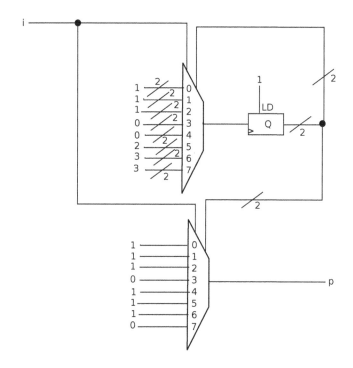

FIGURE 5.6 Circuit diagram from RTL for Example 5.6.

Using our usual approach, we first develop the data path for the RTL description in Example 5.6. We see that the data path only contains two elements: a 2-bit state register, Q, and an output pin, p. These two components are assigned multiple values, and so we will need to use multiplexers to choose between them, as shown in Figure 5.6. In Figure 5.6, one MUX produces the new state for the Q register, and the other MUX produces the output p. The multiplexers are controlled by i, the circuit input, and the current state Q, which are combined into a 3-bit number used as the selector control input to the multiplexers. The data inputs of the multiplexers are connected to the proper constants given in the micro-program for the specified control input. For example, when $iQ_1\overline{Q_0}$ is true, the RTL code specifies that the Q register should be loaded with a value of 3, and the output p should be 1. The minterm $iQ_1\overline{Q_0}$, when read as a 3-bit number, is 110 (decimal 6), and you will see that Option 6 for the MUX attached to the Q register is in fact 3, and Option 6 for the MUX attached to p is in fact 1.

Control for the circuit in Figure 5.6 is relatively simple. The state register, Q, changes state every clock cycle, and so its load line is just wired to a signal of 1. How it changes, is the responsibility of the MUX attached to Q.

Now, what is interesting about this circuit construction is the utter lack of semantic information it contains. The diagram in Figure 5.6 bears no resemblance to the diagram in Figure 3.31. The circuit in Figure 3.31 is the circuit we, presumably, began with in this exercise. It shows the changes to the state, and the output as a Boolean calculation. The diagram in Figure 5.6 shows the change to the state and output, as a seemingly meaningless loading of a choice of constants.

The loss of semantic information is quite common in this type of activity, which could be considered a type of *reverse engineering*. Reverse engineering is the process of generating a high-level description from a low-level description. It is considered to be fairly difficult to automate this process. The problem is not that a low-level description cannot be reverse engineered to a high-level description, but that the results, as in our case, tend to be void of semantics, and, as such, we lose the information on design decisions. Knowing how an artifact was originally designed is important if we wish to intelligently work with the artifact in the future.

5.2   COMMON PROCESSOR MICRO-INSTRUCTIONS

We now have described a tool for specifying circuitry, behaviorally. Behavioral descriptions are usually considered higher level than structural descriptions. The tool, RTL, then allows us to more easily describe circuitry. When we say that we can describe circuitry more easily, what we mean is that our descriptions can be more abstract, with less attention to detail, and using structures that hide complexity.

Because of the higher level of abstraction associated with RTL, it is better at describing large complex circuitry, such as much of the circuitry found in a computer processor. However, it is not only complex circuitry that lends itself to a behavioral description. Low-level circuitry can also be described using RTL. It is, however, in the specification of high-level circuitry where RTL really excels.

Let us begin this section by examining some of the circuits with which we are already familiar, and show how RTL can be used to describe their behavior. In this discussion we will consider both lower-level and higher-level circuitry, demonstrating the flexibility of RTL.

5.2.1   RTL Descriptions of Combinational Circuits

We begin our discussion by considering combinational circuits. In particular, we start with the 4-1 MUX, described in Chapter 3. The interface for the 4-1 MUX is shown in Figure 3.11. Its operation is described by the truth table in Table 3.4. Figure 3.11 shows that the MUX has four single-bit input pins, i_0, i_1, i_2, and i_3, a single-bit output pin, p, and a 2-bit selector input, s. From Table 3.4 we can summarize the behavior of the MUX, by saying that if the selector input is k, then the output, p, would be set to i_k. In RTL this would give us the following micro-program.

$$\begin{aligned}
\overline{s_1} \cdot \overline{s_0} &: p \leftarrow i_0 \\
\overline{s_1} s_0 &: p \leftarrow i_1 \\
s_1 \overline{s_0} &: p \leftarrow i_2 \\
s_1 s_0 &: p \leftarrow i_3
\end{aligned} \tag{5.7}$$

The first micro-instruction in Example 5.7 is executed if the selector input is 0, the second instruction is executed if the selector input is 1, the third if the selector input is 2, and the fourth if the selector input is 3.

5.2.2   RTL Descriptions of Sequential Circuits

Moving on to sequential circuits, the simplest sequential circuit is the flip-flop. We can build an RTL description for the J-K flip-flop. Remember the definition of the J-K flip-flop. Its interface is given in Figure 3.23, and its behavior is given in the excitation table, Table 3.9. The flip-flop has two control inputs, J and K, and an output Q, which is the current state of the flip-flop.

Remember that the J-K flip-flop performs four operations. Which operation is performed is determined by the control lines J and K: lock, when $J = K = 0$; set, when $J = 1$ and $K = 0$; reset, when $J = 0$ and $K = 1$; and compliment, when $J = K = 1$.

We can build the micro-program for the J-K flip-flop directly from Table 3.9. The result is the following micro-code.

$$\overline{J}K : Q \leftarrow 0$$
$$J\overline{K} : Q \leftarrow 1 \qquad\qquad (5.8)$$
$$JK : Q \leftarrow \overline{Q}$$

Notice several things about this RTL description. The three micro-instructions in Example 5.8 correspond to the last three rows of the excitation table. The first two rows have been eliminated from the RTL description. The first row of Table 3.9 shows what happens between clock pulses. A micro-instruction shows what happens at the clock pulse, and it is assumed that the circuit is locked between clock pulses, as is the case with any standard sequential circuit. The second row of Table 3.9 corresponds to a lock operation, in which nothing happens to the state. It is conventional, in RTL, to show only micro-instructions for cases with state change, and it is usually assumed that if a case is not covered by the micro-program, there is no state change in that case.

As a last example, we consider the counter presented in Chapter 3. The interface for the 4-bit counter is shown in Figure 3.39. The counter shown has the control input, IN, the state output, Q, and the carry-out, C_{out}. The input IN specifies a lock operation if it is 0, and an increment operation if it is 1. This behavior can be specified with the following RTL description.

$$IN : Q \leftarrow Q + 1 \qquad\qquad (5.9)$$

You will probably observe how sparse a description we have of the counter. The counter is a fairly complex circuit, and yet the RTL description is fairly simple. This is a graphic demonstration of the power of abstraction of RTL.

5.2.3 Processor Micro-Operations

Hopefully, you are convinced now that RTL is powerful enough to describe most digital circuitry, and that its power of abstraction makes it an effective tool for both high-, and low-level circuitry. At this point we might become interested in what types of micro-operations we find implemented in a normal processor. So far, we have used micro-operations involved in the operation of the standard register: micro-operations for load, micro-operations for increment, and micro-operations for clear. These micro-operations form a rather small vocabulary, and so now we expand that vocabulary by introducing more types of micro-operations.

5.2.3.1 *Arithmetic Micro-Operations*

We begin with micro-operations concerned with arithmetic.

1. Addition: $X \leftarrow X + Y$

2. Subtraction: $X \leftarrow X - Y$

3. Increment: $X \leftarrow X + 1$

4. Decrement: $X \leftarrow X - 1$

5. Transfer: $X \leftarrow Y$

6. Clear: $X \leftarrow 0$

These micro-operations perform the usual addition operations. Operation 5, called transfer, is nothing more than our load operation. Operation 6 is our usual clear operation. Operations 5 and 6 are listed as arithmetic operations, because, as we shall see when we start designing the processor, it is convenient to think of these operations as addition, with all or one of their operands zero.

5.2.3.2 Logic Micro-Operations

Another type of micro-operation is the logic operation. These operations are concerned with implementing Boolean operators. The common logic operations are given below.

7. AND: $X \leftarrow X \wedge Y$

8. OR: $X \leftarrow X \vee Y$

9. NOT: $X \leftarrow \overline{X}$

10. XOR: $X \leftarrow X \oplus Y$

On the surface, these look like the standard Boolean operations, until you realize that their operands are multi-bit. For example, each of the registers, X, and Y might be 8-bit registers. And, the question then arises as to what the result is when performing an AND operation with two 8-bit operands.

To work with multi-bit operands, processors typically perform *bitwise* Boolean operations. Bitwise operations perform the operation on the corresponding bits of the two operands, without consideration of the other bits. As an example, suppose that you are ANDing two 4-bit values; $0110 \wedge 0101$. To perform a bitwise AND, you first AND the Position 0 bits (the low-order bits), independently of the other bits, with the result of $0 \wedge 1 = 0$. You then AND the Position 1 bits, $1 \wedge 0 = 0$, the Position 2 bits, $1 \wedge 1 = 1$, and finally the Position 3 bits (the high-order bits), with the result $0 \wedge 0 = 0$. The full result is then produced by assembling the bitwise results and arranging them, right to left, in lowest to highest bit order: 0100. If we arrange the problem vertically, you will see that we are simply applying the Boolean operator to each column, as shown below.

$$
\begin{array}{c}
0110 \\
\underline{\wedge\ 0101} \\
0100
\end{array}
\qquad \Rightarrow \qquad
\begin{array}{cccc}
0 & 1 & 1 & 0 \\
\underline{\wedge\ 0} & \underline{\wedge\ 1} & \underline{\wedge\ 0} & \underline{\wedge\ 1} \\
0 & 1 & 0 & 0
\end{array}
$$

Notice that in Micro-operations 7 and 8, for the Boolean operators AND and OR, we have changed our usual notation. We normally use the plus sign to indicate OR, and the dot to indicate AND. In micro-instructions, we switch to the symbols $\wedge$ and $\vee$. This is because we cannot use the plus sign as the OR operator, since we are already using it to denote addition. That is, if we see $X \leftarrow X + Y$, an addition operation is being performed. So, to denote the OR operation we need a new notation.

5.2.3.3 Shift Micro-Operations

The next category of micro-operations is that of the shift operations. Shift operations perform the same type of task as is done by a shift register. They move the bits of an operand either to the left, or to the right.

11. Logic Shift left: $X \leftarrow \text{shl } X$

12. Logic Shift right: $X \leftarrow \text{shr } X$

13. Circular shift left: $X \leftarrow \text{cir } X$

14. Circular shift right: $X \leftarrow \text{cil } X$

15. Arithmetic shift left: $X \leftarrow \text{ash} l\ X$

16. Arithmetic shift right: $X \leftarrow \text{ashr } X$

Remember that we introduced the concepts of shifting in Section 3.2.5.2. In Figure 3.37 we show both the conceptual view of a shift-left and a shift-right. Recall that we introduced the concepts of a carry-in and a carry-out. The three types of shifts, given in Operation 11 through Operation 16, are all variations of this same shift pattern, that differ only in how the carry-in and carry-out are handled.

The logical shift is a shift in which $c_{in} = 0$. That is to say, the vacated bit is filled with 0. For example, consider the 4-bit value 0110. If we were to shift it left, each bit would move one bit up, and the low-order bit would be filled with 0, resulting in 1100. If we were to shift it right, each bit would move down one bit, and the high-order bit would be filled with 0, resulting in 0011.

The circular shift, also called a rotate, is defined by the rule $c_{in} = c_{out}$. That is to say, the vacated bit is filled with the bit that falls off the other end. Another way of saying this is that the bit that falls off the end of the value wraps around, and comes back in to fill the vacant spot. If we shift the 4-bit value 1010 left, with a circular shift, all bits move up one position. The high-order 1 drops off, and wraps around to fill the low-order bit, resulting in the value 0101. If 1010 is shifted right, all bits move down. The rightmost 0 drops off, and wraps around to fill the high-order slot, resulting in the same value, 0101.

The arithmetic shift must be considered in two separate cases, depending on whether you are shifting left or right. For a left shift $c_{in} = 0$. You will observe that the arithmetic shift left is the same as the logical shift left. The difference is only in the name. The names of the shifts reflect their usage, rather than any actual operational difference.

The name *arithmetic shift* emphasizes that a shift can be, and is being used to perform arithmetic. Shifting left is an easy way to perform multiplication. For example, in decimal, if we take the number 3, and shift it left one digit, we get 30. This is the result of multiplying the number 3 by the base of the number system, 10. If we then shift 3 left again, we get 300, which is the result of multiplying 3 by 10, twice.

In general, left shifts can be thought of as multiplying numbers by powers of the base. This extends to binary. For example, if we take the 8-bit number 00000011 (decimal 3) and shift it left, we get 00000110 (decimal 6), which is the result of multiplying the 3 by the base 2. Shifting again gives us 00001100 (decimal 12), which is the result of multiplying by 2, again.

The arithmetic right shift represents division by 2, and is defined by $c_{in} = sign$. As discussed in Chapter 2, the *sign* bit is the high-order bit of a binary number. To illustrate the workings of right shifts, let's look at a couple of examples. First, consider the 8-bit number 00010111 (decimal 23). If we shift this number to the right we get 00001011. For now, we have just used 0 as the carry-in, resulting in decimal 11. If we are doing integer division, 11 is, in fact, the result of dividing 23 by 2. (Remember that if we divide 23 by 2, using integer division, the result is $\lfloor 23/2 \rfloor = 11$.

As a second example, consider the 8-bit number 11101001. This number, with sign bit 1, is a negative number (-23 in decimal). If we shift this number right, and use 0 as the carry-in, the result is 01110100 (decimal $+116$). This result does not represent $\lfloor -23/2 \rfloor = -12$. It is not even negative, as it should be! However, if we use 1 as the carry-in, instead of 0, we get 11110100, which is, in fact, -12 in decimal. What we have learned is that if we desire a right shift that represents division by 2, for negative numbers the carry-in must be 1, and for non-negative numbers the carry-in must be 0. An equivalent, but more succinct rule is $c_{in} = sign$, as previously stated. That is to say, when shifting right, all bits are moved down. This includes the high-order bit, which is moved down one bit, but also copied back into the vacant spot at the left.

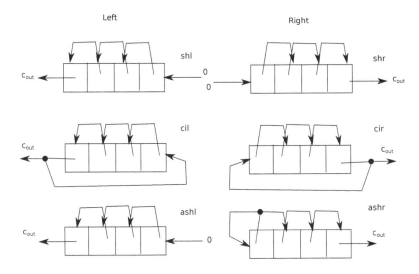

FIGURE 5.7 Illustration of the three shift types.

In Figure 5.7 we summarize the three types of shifts. The logical shift shows a carry-in of 0. The circular shift shows the carry-out routed to the carry-in. The arithmetic shift shows a carry-in of 0 for the left shift, and the sign bit as the carry-in for the right shift.

5.2.3.4 Memory Access Micro-Operations

So far, all of our micro-operations have been operations on registers located inside the processor. We now turn to a couple of operations used to transfer information in and out of the processor. In particular, these operations are used to send and receive information from and to the memory unit.

Remember from Chapter 4 that memory units typically have a read operation and a write operation. From the processor's perspective, a read operation results in the value of a register being loaded in from the memory unit. In a write operation, the value of a register is stored to a particular memory location. These two operations, in RTL, are shown below.

17. Read: $X \leftarrow M[AR]$

18. Write: $M[AR] \leftarrow X$

In the read operation, Operation 17, the register X receives the contents of the memory location, with the address given by the register AR (address register). We use this register name to emphasize a common design decision in processor design. Typically the address bus is hardwired to a single register, the AR. The result of this decision is that if you wish to address the memory unit, you must place the address in the AR register. In other words, if you wish to perform the micro-operation $X \leftarrow M[PC]$, it must actually be performed as two micro-instructions:

$$AR \leftarrow PC$$
$$X \leftarrow M[AR]$$

This design decision, versus allowing any register to be used for addressing, can slow the processor a bit, but it does have an important advantage. In particular, it simplifies the design of the address bus.

The write operation, Operation 18, stores the contents of the register X into the memory location addressed by the register AR.

5.3 ALGORITHMIC MACHINES

Most people cannot help but notice the similarity between a micro-program, written in RTL, and an actual program, written in a high-level programming language. We have seen that RTL is a *declarative language*, describing how a circuit is put together, rather than a *procedural language*, which gives a sequence of executable steps that constitute an algorithm. The fact that RTL is not procedural, however, does not mean that we cannot use it to describe algorithms.

You may have noticed that many hardware machines can be thought of in algorithmic terms. For an example, a cruise control on an automobile can be thought of as a program, performing a sequence of steps, including checking various sensors and adjusting a throttle position. In fact, the cruise control may actually contain a processor that is executing software that does exactly this kind of control. On the other hand, the cruise control unit might very well be a solid piece of hardware, and it is the hardware that performs the steps to control the speed.

Let's suppose that we have an algorithm that we wish to implement as a machine. We see that we have a choice in how the algorithm is implemented.

We can implement it as pure hardware, circuitry that performs the algorithmic steps. We can implement it as pure software, and run the software on a general-purpose computer. Or, we can compromise, and build parts as hardware and other parts as software.

In the above discussion, where we say we have a choice as to whether an algorithm is implemented as hardware or software, we are assuming that we know how to turn an algorithm into hardware. This is a topic we have not covered yet, and which we now address.

5.3.1 The Teapot Example

Probably the best way to show how to turn an algorithm into a circuit is with an example. Suppose that you are interested in building a machine to control an electric tea pot. The tea pot has an on-off switch. A sensor on the switch produces a signal, S, where $S = 1$ if the pot is on, and $S = 0$ if the pot is off. A sensor detects the temperature of the liquid in the pot. It produces a signal T, where $T = 1$, if the liquid is hot enough, and $T = 0$ if the liquid is too cold.

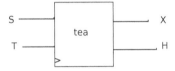

FIGURE 5.8 Interface for tea pot.

The control circuit you are building will regulate the tea pot. The interface to the control circuit is shown in Figure 5.8. The inputs to the controller are the signals from the two sensors, S and T, and the clock. The outputs of the controller are two signals, H and X. When H is set, the heating element in the pot is turned on. When H is clear, the heating element is turned off. X is a signal that turns off the on-off switch, when set to a 1. In a typical use-case scenario, the user fills the pot with water, turns on the switch, and waits until the switch flips to the off position again, indicating that the water is hot enough.

You have put a lot of thought into the circuit design, and realize that the switch might be activated by an object leaning against the tea pot. In this case, when the switch is turned off by the control circuit, the pot will remain on, with the switch stuck in the on position. As a consequence you decide that the pot should not go on, after a successful brew, until the sensor S shows that the switch has returned to the off position.

Our first task in building this circuit might be to write an algorithm that describes the functioning of the control circuit. That algorithm might resemble the following code.

```
stuck = 0
loopforever
    if S and not stuck and not T then
        H = 1
        X = 0
        stuck = 0
    else if S and not stuck and T then
        H = 0
        X = 1
        stuck = 1
    else if S and stuck then
        H = 0
        X = 0
        stuck = 1
    else if not S then
        H = 0
        X = 0
        stuck = 0
```

$$(5.10)$$

This code executes the contents of the *loopforever* structure every clock cycle. A Boolean variable *stuck* is used to determine if the switch is stuck in the on position. Inside the loop, a case statement checks the sensor inputs and the *stuck* variable. The top case indicates that the pot is heating normally. In this case the heat is turned on, and there are no changes to the output X and the variable *stuck*. The next case is when the pot has heated to the required temperature. The heating element is then turned off, the on-off switch is turned off, and the variable *stuck* is set to the value true, to cause the system to wait until it detects that the switch is actually off. In the third case, we see the situation where the system is stuck, and the switch has not gone off yet. In this case the heat is turned off, and the system remains in the stuck state. The last case is when the on-off switch is in the off position, in which case the heat is set to off, and the variable *stuck* is cleared.

5.3.2 Generating a Flowchart, and the Role of the Sequencer

Once we have a working algorithm, we can start the process of converting it into RTL. A useful exercise, which helps us on the way to RTL, is to draw a flowchart for the algorithm. Our flowchart is shown in Figure 5.9. The flowchart is composed of decision blocks and action blocks. You will notice

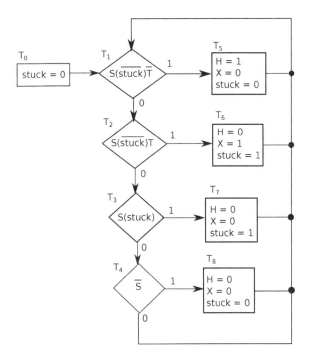

FIGURE 5.9 Flowchart for tea pot, Example 5.10.

that we have given each block a name. For instance, the first block to the left is named T_0. What we are emphasizing with these names is that the flowchart explicitly shows a sequence of steps, each step performed at a different time. The name T_0 indicates that we might think of this action as being performed at Time Step 0. At Time Step 1 we are performing the first decision, and so on.

Having actions performed at discrete time steps is one of the important characteristics of the algorithm, made clear by the flowchart. When the algorithm is turned into circuitry, each of the discrete time steps can be implemented as a micro-instruction, performed in one clock cycle. Another important characteristic, observed from the flowchart, is the similarity of the flowchart to an FSM. We can think of the blocks as states, and the flow arrows as state transitions. This observation suggests that the circuit we build might be constructed with similar structure to the sequential circuits we build from FSMs. That is to say, we might build a circuit that has a state register, storing the current state, and control circuitry to calculate the next state, and the output.

The state register for an algorithmic circuit can be called a *sequencer*. Its job is to count through the time steps for the algorithm. The sequencer usually stores the time step number. So, for example, when the current time step is T_4, the sequencer would contain a 4. When the current time step is T_2, the sequencer would contain a 2.

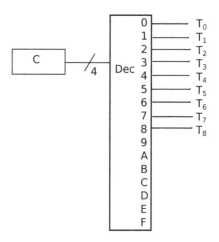

FIGURE 5.10 Sequencer for the tea pot example of Figure 5.9.

The number stored in the sequencer is typically decoded into a trigger signal. The structure to do this, for the tea pot example, is shown in Figure 5.10. The sequencer is the register C. It is a 4-bit register. The four bits are required in order to store numbers between 0 and 8, that correspond to the time steps T_0 to T_8. The four bits of the sequencer are fed into a 4-16 decoder. Of the 16 output lines, only the first eight are used as trigger lines in the micro-program circuit.

5.3.3 Generating RTL from the Flowchart

The sequencer is the mechanism we use to represent the concept of time in an algorithm. Now that we have this mechanism, we can start writing the RTL micro-instructions. We write one or more micro-instructions for each time step. Starting with Time Step 0, we see that the variable *stuck* is set to a 0. Variables in the algorithm are implemented using registers. So, we would have a register in our circuit called *stuck*, and Time Step 0 would be implemented with the micro-instruction

$$T_0 : stuck \leftarrow 0, C \leftarrow 1.$$

This micro-instruction specifies that *stuck* is set to 0, and that the sequencer value should be changed to 1, indicating that the next time step, or state, is T_1.

Moving on to Time Step 1, we see that no action is performed at this time. The result of this time step is to either set the sequencer to 0 or 2, corresponding to the time steps T_0 or T_2. This time step will generate two micro-instructions.

$$\frac{T_1 S(stuck) \cdot \overline{T} : C \leftarrow 5}{T_1 S(stuck) \cdot \overline{T} : C \leftarrow 2}$$

Each micro-instruction specifies the action to perform for one of the alternate decision branches; one if the condition is true, and one micro-instruction if the condition is false.

If we continue processing the states of Figure 5.9, we would arrive at a micro-program like the following.

$$
\begin{aligned}
&\textbf{Def} : T_0 \equiv C = 0, T_1 \equiv C = 1, T_2 \equiv C = 2, T_3 \equiv C = 3, \\
&\quad T_4 \equiv C = 4, T_5 \equiv C = 5, T_6 \equiv C = 6, \\
&\quad T_7 \equiv C = 7, T_8 \equiv C = 8 \\
&T_0 : stuck \leftarrow 0, C \leftarrow 1 \\
&T_1 S(stuck) \cdot \overline{T} : C \leftarrow 5 \\
&T_1 \overline{S(stuck) \cdot \overline{T}} : C \leftarrow 2 \\
&T_2 S(stuck)T : C \leftarrow 6 \\
&T_2 \overline{S(stuck)T} : C \leftarrow 3 \\
&T_3 S(stuck) : C \leftarrow 7 \\
&T_3 \overline{S(stuck)} : C \leftarrow 4 \\
&T_4 \overline{S} : C \leftarrow 8 \\
&T_4 S : C \leftarrow 1 \\
&T_5 : H \leftarrow 1, X \leftarrow 0, stuck \leftarrow 0, C \leftarrow 1 \\
&T_6 : H \leftarrow 0, X \leftarrow 1, stuck \leftarrow 1, C \leftarrow 1 \\
&T_7 : H \leftarrow 0, X \leftarrow 0, stuck \leftarrow 1, C \leftarrow 1 \\
&T_8 : H \leftarrow 0, X \leftarrow 0, stuck \leftarrow 0, C \leftarrow 1
\end{aligned}
\tag{5.11}
$$

The micro-program starts with definitions of the time step trigger signals. For instance, T_2 is defined to be a 1 if and only if $C = 2$. Following the trigger signal definitions, the micro-program proceeds with micro-instructions for the states in Figure 5.9, derived in the same way that we derived micro-instructions for the states T_0 and T_1.

Our conversion of this algorithm is a good demonstration of how to convert an algorithm into a behavioral description. The process we have followed is mechanical. Although the process yields a working circuit, it does not yield a very efficient circuit. This mechanical conversion yields a system that reacts to sensor changes in at most five clock cycles. It is possible to rewrite the circuit to react in a single clock cycle, but optimality was not our goal. Our goal was to describe a mechanical transformation.

We have now successfully converted an algorithm into a micro-program. This micro-program can be further converted into a circuit. This is a task we have performed several times already, manually. The problem with a manual conversion is that this RTL micro-program is a little more complex than the examples we worked with previously. And, we must acknowledge that eventually RTL micro-programs will become too complex for easy manual conversion. Fortunately the process of circuit generation from a behavioral description has been automated.

5.4   RTL AND VERILOG

We use RTL to describe circuitry, in a behavioral fashion. This is useful, since behavioral descriptions can be higher level, and more abstract than structural descriptions. What would be more useful is if, not only could we describe a circuit with RTL, but that once the circuit is described, we could get the circuit built for us automatically. The problem is that to succeed at building a program to generate circuitry, we need a much more precise language than RTL to work with.

The problem is that, when using RTL as a design tool, we usually leave our descriptions incomplete and ambiguous. For example, in Example 5.11, we do not specify the size of the register C, a critical piece of information, if RTL is to be used for circuit construction. RTL is designed for flexible use by circuit designers, and not so much for use by automating software. An actual hardware description language, geared towards automation, is often, in fact, harder to use in design, because the designer must worry about detail, and syntax, which diverts attention from the design process.

If you compare the situation with programming languages, RTL is very much like pseudo-code. Pseudo-code is a rather informal language intended for use in software design, and not intended for execution on a computer. In order for a programmer to implement the software, the pseudo-code must be manually translated into a good production language, like C++ or Java. Pseudo-code is designed to be easy to translate into a high-level language, which makes it a good choice for design; not only does it relieve the programmer from detail during the design process, but when it comes time to implement the code, it is easily translated into a more precise language.

If we wish to automate circuit generation from RTL, we need a more formal version of RTL, for a circuit compiler to work with. We need the equivalent of a language like C++ for circuit specification. One such more formal version of RTL is a language called Verilog.[1]

In this section we show how an RTL program might be rewritten in Verilog. This section is not a Verilog reference or a Verilog tutorial. It is meant merely to briefly demonstrate the relationship between RTL and Verilog.

```verilog
// tea pot controller                        1
module teapot(clk, S, T, H, X);              2
  // input ports                             3
  input clk, S, T;                           4
  // output ports                            5
  output reg H, X;                           6
  // internal registers                      7
  reg stuck;                                 8
  reg [3:0] C;                               9
  // define the states                       10
  assign T0 = C == 4'b0000;                  11
  assign T1 = C == 4'b0001;                  12
  assign T2 = C == 4'b0010;                  13
  assign T3 = C == 4'b0011;                  14
  assign T4 = C == 4'b0100;                  15
  assign T5 = C == 4'b0101;                  16
  assign T6 = C == 4'b0110;                  17
  assign T7 = C == 4'b0111;                  18
  assign T8 = C == 4'b1000;                  19
  // the circuit behavior                     20
  always @(posedge clk) begin                21
    if (T0) begin                            22
      stuck = 0;                             23
      C = 4'b0001;                           24
    end                                      25
    if (T1) begin                            26
      if (S && !stuck && !T)                 27
        C = 4'b0101;                         28
```

---

[1] Just as with programming languages, where C++ is not the only available implementation language, Verilog is not the only language available for hardware description. Another popular language is VHDL (VLSIC Hardware Description Language, where VLSIC is an acronym for Very Large Scale Integrated Circuitry). The acronym HDL is used to describe the type of language, like Verilog, and VHDL, used to describe circuitry. And, in fact, Verilog is often referred to as Verilog HDL.

```
      else                              29
        C = 4'b0010;                    30
    end                                 31
    if (T2) begin                       32
      if (S && !stuck && T)             33
        C = 4'b0110;                    34
      else                              35
        C = 4'b0011;                    36
    end                                 37
    if (T3) begin                       38
      if (S && stuck)                   39
        C = 4'b0111;                    40
      else                              41
        C = 4'b0100;                    42
    end                                 43
    if (T4) begin                       44
      if (!S)                           45
        C = 4'b1000;                    46
      else                              47
        C = 4'b0001;                    48
    end                                 49
    if (T5) begin                       50
      H = 1;                            51
      X = 0;                            52
      stuck = 0;                        53
      C = 4'b0001;                      54
    end                                 55
    if (T6) begin                       56
      H = 0;                            57
      X = 1;                            58
      stuck = 1;                        59
      C = 4'b0001;                      60
    end                                 61
    if (T7) begin                       62
      H = 0;                            63
      X = 0;                            64
      stuck = 1;                        65
      C = 4'b0001;                      66
    end                                 67
    if (T8) begin                       68
      H = 0;                            69
      X = 0;                            70
      stuck = 0;                        71
```

```
        C = 4'b0001;                                    72
    end                                                 73
  end // behavior                                       74
  // initialize the state                               75
  initial begin                                         76
    C = 4'b0000;                                        77
    H = 0;                                              78
    X = 0;                                              79
  end                                                   80
endmodule                                               81
```

Listing 5.1 Verilog code for the teapot example, Example 5.11.

To illustrate the relationship between RTL and Verilog, we present the Verilog code in Listing 5.1, derived from the tea pot example micro-program, Example 5.11. In Verilog, a circuit is described in a *module*. The module is defined by a *module expression*. In Listing 5.1 this expression starts at Line 2, and ends at Line 81. The heading of the *module expression* gives the circuit a name, and defines the inputs and outputs for the circuit, much as we did using the pin-out diagram from Figure 5.8. In Verilog, the inputs, and outputs are called *ports*.

In the *module* expression, we give only the port names. Immediately following the heading of the *module* expression, in Lines 4 and 6 of Listing 5.1, we specify the types for the names. *Clk*, *S*, and *T* are declared input ports, and *H* and *X* are declared output ports. We also declare the registers used in the circuit. *Stuck* is declared as a 1-bit register, and *C* is declared as a 4-bit register. To declare *C* as 4 bits, we specify that its bits are numbered 3 through 0. With 4 bits, *C* is capable of storing state numbers 0 through 8, which are the states of our machine, shown in Figure 5.9.

Next in the program, we define the values T_0 through T_8, just as we did at the start of the RTL in Example 5.11. These declarations are done using the *assign* declaration. This declaration defines a signal based on some computation. In this case the signals are being computed by circuitry comparing *C* with constants, for equality. For instance T_4 is compared with the 4-bit binary constant 0100. In Verilog, the syntax for this constant is 4'b0100, indicating that the constant is 4-bits, binary, and then giving the values of the four bits.

The Verilog code in Listing 5.1 now moves on to the information contained in the body of the RTL micro-program. This information is enclosed inside of an *always* expression, spanning Lines 21 through 74. The *always* expression specifies behavior that is performed when a condition is satisfied. In our example, the condition specifies that the behavior should occur "at" a particular point in time, only. That point in time is given by the pseudo-function *posedge*, which specifies the positive, or rising edge, of the clock signal *clk*.

The micro-instructions, from the RTL code are implemented as a sequence of *if* expressions. Lines 22 through 25 of Listing 5.1 implement the micro-instruction for T_0. The control is specified as the condition of he *if* expression, and the micro-operations are the body of the *if* expression.

Lines 26 through 31 of Listing 5.1 implement the two micro-instructions for the state T_1. The control specifications for these two micro-instructions check not only that the state is T_1, but also whether a second condition, $S\overline{(stuck)} \cdot \overline{T}$, is true or false. In the Verilog code, this second condition is checked with an *if* expression, spanning Lines 27 through 30 of Listing 5.1.

The micro-instructions for states T_2 through T_4 follow, in Lines 32 through 49 of Listing 5.1. The code for these micro-instructions is similar to the code for the state T_1. In Lines 50 through 73 we implement the micro-code for states T_5 through T_8 with *if* expressions, using the condition to do the control, and the body to do the operations.

The last section of the Verilog code, Lines 76 through 80 of Listing 5.1, initializes the registers used by the circuit. This is a job that is ignored when writing RTL. When building a circuit, however, in which registers are used, the registers must have values before the circuit tries to use their current value for control purposes. The *initial* block in Verilog performs a set of assignments at circuit power-up.

Comparing Verilog to RTL we see that Verilog, clearly, has a much more rigid syntax. We also observe that features of the circuit that are implicit in RTL are made explicit in Verilog. The initialization of registers is one example of the implicit becoming explicit. Another example is the type declaration of the output ports. In RTL, H and X are treated as if they are just wires. But, RTL has an implicit assumption that any signal left unchanged, remains the same, until the next clock cycle. But, to do this, an output port must be able to remember its current value. This cannot be done by just a wire; it requires a register. Notice that our outputs have a declared type *output reg*, in Line 6 of Listing 5.1, indicating that the output line is connected to a register, and it is the register that is actually being manipulated when the port is manipulated.

5.5 SUMMARY

We have described register transfer language in this chapter. RTL is a tool used to describe circuitry. It can be used to describe circuitry at any level, although it excels at describing higher-level circuitry.

We started the chapter by describing how RTL descriptions can be converted into custom circuitry. We also spent some time explaining how sequential machines, given in the form of an FSM, can be converted into RTL descriptions. Combined, these two discussions give us the information required to convert from structural specification, to behavioral specification, and then to circuit implementation.

This chapter then moved on to talk about some of the more common micro-operations used in processor design. These operations can be classified as arithmetic, logic, shift, and memory access operations.

We included a section demonstrating how to convert an algorithm into a circuit. The technique uses a sequencer to introduce timing. With the ability to convert an algorithm into hardware, we now have a choice, when designing a machine. Algorithms can be implemented as software, which is a fairly typical scenario, as hardware, using the technique discussed, or as a combination of hardware and software.

We finished the chapter by comparing RTL to Verilog HDL. RTL serves as a pseudo-code version of Verilog. It allows the designer to design a circuit, free of strict syntactic constraints, and detailed declaration. When it comes time to convert the RTL to actual Verilog code, however, the conversion is relatively easy.

With our newly acquired command of RTL, we are almost ready to design a processor. One last discussion, though, remains; we need to talk more about what exactly a processor does. But, in terms of design tools, at this point we have acquired almost all the information needed.

TABLE 5.2 State transition table for Exercise 5.3.

j	$Q_{(0)}$	$Q_{(1)}$	v
0	0	0	1
0	1	1	0
1	0	1	1
1	1	0	1

5.6 EXERCISES

5.1 Draw a circuit schematic for the following RTL circuit description.

$$\bar{s}\bar{t} : R1 \leftarrow R1 + R3, R2 \leftarrow R2 + 1$$
$$\bar{s}t : R3 \leftarrow 0$$
$$s\bar{t} : R3 \leftarrow R1, R2 \leftarrow R1 \qquad \text{(P5.1)}$$
$$st : R3 \leftarrow R2 + R3$$

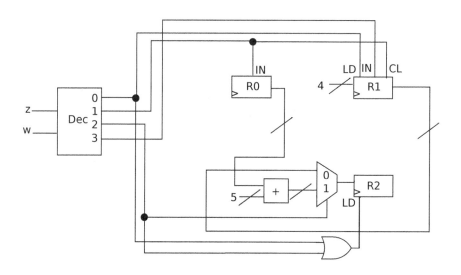

FIGURE 5.11 Circuit for conversion to RTL in Exercise 5.2.

5.2 Analyze the circuit diagram of Figure 5.11, by giving a micro-program that describes the circuit.

5.3 Given the transition table in Table 5.2,

    a. give micro-instructions describing the circuit and

b. draw a schematic of the circuit, from the micro-instructions in your answer to Part 5.3a.

5.4 Build a machine with two 4-bit registers, X and Y, that performs the following operations.

$$\overline{OP_1} \cdot \overline{OP_0} : X \leftarrow X + Y$$
$$\overline{OP_1}(OP_0) : X \leftarrow X - 1$$
$$(OP_1)\overline{OP_0} : X \leftarrow X + 1 \qquad \text{(P5.2)}$$
$$(OP_1)(OP_0) : X \leftarrow Y$$

(Note: you can perform the micro-operation $X \leftarrow X - 1$ by performing the micro operation $X \leftarrow X + (-1)$. Remember that a 4-bit -1 is the two's compliment of the 4-bit number 0001, which is 1111.)

5.5 Build machines with two 4-bit registers, X and Y, that perform the following bit-wise operations. Draw diagrams similar to Figure 3.38.

a. $(EN) : X \leftarrow X \wedge Y$

b. $(EN) : X \leftarrow X \vee Y$

c. $(EN) : X \leftarrow \overline{X}$

d. $(EN) : X \leftarrow X \oplus Y$

5.6 Draw the following 4-bit shift registers.

a. $(SH) : Q \leftarrow \text{shr } Q, C_{out} \leftarrow Q_0$

b. $(SH) : Q \leftarrow \text{cil } Q, C_{out} \leftarrow Q_3$

c. $(SH) : Q \leftarrow \text{ashr } Q, C_{out} \leftarrow Q_0$

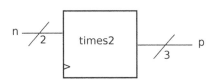

FIGURE 5.12 Interface for *times2* circuit used in Exercise 5.7.

5.7 You are designing a machine that accepts an input n, and produces an output p. The interface for the machine is shown in Figure 5.12. The algorithm follows.

```
loopforever
    z = 0
    x = n
    while not (x = 0) do                    (P5.3)
        x = x - 1
        z = z + 2
    p = z
```

a. Draw a flowchart for the algorithm.

b. Write the RTL micro-program for the algorithm.

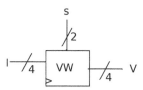

FIGURE 5.13 Interface for VW circuit of Exercise 5.8.

5.8 Write a Verilog program for the following RTL micro-code. The interface is given in Figure 5.13. Test your code on a Verilog simulator.

$$\overline{s_1} \cdot \overline{s_0} : W \leftarrow I, V \leftarrow W$$
$$\overline{s_1} s_0 : V \leftarrow 0 \qquad\qquad (P5.4)$$
$$s_1 \overline{s_0} : W \leftarrow W + 1$$

5.9 Rewrite your RTL description from Exercise 5.7 as Verilog code. Test your code on a Verilog simulator.

Common Computer Architectures

CONTENTS

I N CHAPTER 3 we have explained that a processor is a device that repeatedly executes the machine cycle, in which it fetches an instruction from memory, decodes the instruction, and executes it. The instructions being fetched from memory are part of a program. In Chapter 1 we discussed how the program ends up in memory, involving the processes of compilation, assembly, linking, and loading. What we don't know yet is what sort of operations a running program can perform. Our situation is a bit like knowing the grammar rules for a foreign language, but not knowing any vocabulary.

In this chapter we explore what sort of instructions a computer can execute. We also look at how these instructions are represented numerically. (Recall that machine language, the actual native language of a computer, is numeric.) These topics are usually classified as topics in the field of *computer architecture*.

When we start looking at ways to organize processors, we find that there are many ways to do this; some of them good, some of them better, and some of them not so good. But, if we look at the computer architectures that are useful, we find that there are several, and we find that these several architectures are all used in current processor designs. So, this chapter is also a survey of some of the different architectures in use today.

6.1   INSTRUCTION SET ARCHITECTURE

When we use the term *Instruction Set Architecture* (ISA) we are referring to the set of instructions a processor is capable of executing, and the form of the instructions. The instructions in the instruction set are machine language instructions. Every different processor family has its own particular machine language, making machine language non-portable.

The fact that machine language is numeric makes it somewhat difficult to work with, at least for human users. Fortunately, as explained in Chapter 1, machine languages have symbolic versions, called assembly languages. In assembly language we assign names to the numeric components of the machine language. Assembly languages are much easier to work with, and so it is not surprising that we will begin our discussion by studying assembly language. It is, however, important that we remember that we are primarily interested in examining machine language, and that we are simply using assembly language as a tool to facilitate this.

Probably the best way to develop an understanding of ISA topics is through an example. So, we start our discussion by describing a processor from the machine language level.

The user of a machine writes programs for the machine. These programs are a collection of instructions, to be executed by the machine. The instructions are drawn from a set of instructions which the processor understands, and can execute. The term *instruction set architecture* emphasizes that the set of instructions available on a machine depends on the architecture, or structure, of the machine.

Regardless of machine architecture, we can categorize the instructions available into the following categories.

- Data transfer.
- Data manipulation.
- Control.

6.1.1 Data Transfer

Data transfer operations move data from one location to another. We can further categorize data transfer instructions into the following sub-categories, based on the location of the operands.

- Register-to-Register.
- Register-to-Memory.
- Memory-to-Register.
- Register-to-Device.
- Device-to-Register.

6.1.1.1 Register-to-Register Transfer

Register-to-register instructions copy the data in one register to another register. In assembly language, a register-to-register instruction might be similar to the following.

```
mov R0, R1
```

This instruction moves the contents of Register 1 to Register 0. Or, if we defined the meaning of the instruction in RTL, we would get $R0 \leftarrow R1$.

6.1.1.2 Register-to-Memory Transfer

Register-to-memory instructions copy the value of a register to a memory location. An example would be the following.

```
store 5, R1
```

An instruction that transfers data from the processor to memory is typically called a *store* instruction, as opposed to a *mov* instruction, that performs a transfer that is internal to the processor. Our example instruction would be defined, in RTL, as $M[5] \leftarrow R1$.

6.1.1.3 Memory-to-Register Transfer

A memory-to-register instruction does the opposite of a register-to-memory instruction, copying from a memory location to a register. The verb we use to specify a memory-register transfer is *load*, as in the following assembly example.

```
load R0, 5
```

This instruction would be defined in RTL as $R0 \leftarrow M[5]$.

6.1.1.4 Device Transfer

Moving on to the device movement instructions, device-to-register, and register-to-device instructions, these are instructions that move data to and from peripheral devices.

In general a machine may have several devices connected to it. As a consequence, I/O instructions must designate to, or from which device the transfer is occurring. This is done by numbering each device with a unique index, called a *channel number*. These channel numbers are then used in the transfer instruction.

Moving data to an output device on Channel 3 is done using the following type of instruction.

```
out 3, R0
```

This instruction is defined as performing the operation $D_{out,3} \leftarrow R0$, where $D_{out,3}$ is the device register of the device on Channel 3. Receiving data from an input device would use an instruction similar to the following.

```
in R0, 3
```

This instruction performs the operation $R0 \leftarrow D_{in,3}$, where $D_{in,3}$ is the value coming from the device on Channel 3.

It is worth mentioning that there are two I/O architectures in common usage: *memory-mapped I/O* and *special instruction I/O*. On machines that implement memory-mapped I/O, devices are mapped to a logical memory location. For example, the memory location M[255] might be an output device.

That is to say, when the user places a value in this location, it is sent to the output device. So, to output the value in R0, you would simply use the **store** instruction, as follows.

```
store 255, R0
```

In special instruction I/O, special instructions exist for doing I/O. We use the *in* instruction to collect input from an input device, and the **out** instruction to dispatch output to an output device.[1]

6.1.2 Data Manipulation Instructions

Data manipulation instructions perform operations on data. On many machines, it is required that data being operated on is in registers. That is to say, data from memory or devices must be moved into registers before it can be processed. This requirement helps speed up data manipulation, since the data manipulation instructions are not burdened with memory read operations and memory write operations, in addition to their manipulation operation.

6.1.2.1 Common Data-Types

In order to understand data manipulation, we need to understand, a bit more, what data is. Computers work on various types of data. We list some of the more common data-types.

- Integer data.

- Real data.

- Boolean data.

- Character data.

- Binary coded decimal (BCD) data.

6.1.2.2 The Integer Data-Type

The integer data-type is used to represent whole numbers, meaning numbers that have no non-zero fractional part. As discussed in Chapter 4, the word size of a computer is usually geared towards the integer type, and as such, usually an integer is stored in a single computer word. The bits of the word form a binary number which represents the integer value.

---

[1]It should be noted that even if a machine does special instruction I/O, if it uses a bus architecture, the I/O device is still mapped to a logical memory location via its bus address, in a way which can be considered memory mapping. So, you can consider an *out* instruction as a shorthand for a *store* operation to that location in logical memory, and an *in* instruction as a *load* from that location.

Computers usually support two integer types: unsigned integers and signed integers. The unsigned representation is used to represent non-negative numbers only, and all bit configurations of the computer word are used to represent numbers greater than or equal to zero. In the signed representation, both negative and non-negative integers are represented. As such, half of the word bit configurations are used to store negative values, and the other half of the configurations are used to represent non-negative values.

Machine word sizes must be large enough to accommodate the sizes of integers that users wish to operate on. A word size of 8 bits can represent integers from 0 (00000000 in binary) to $255 = 2^8 - 1$ (11111111 in binary). The value 255 is typically considered a small integer, and users often work with much larger integers. Because of this you often see computer word sizes of 32 bits, or even 64 bits. With a 32-bit word, the highest number that can be represented is $2^{32} - 1 > 4 \times 10^9$, which obviously gives the user a much larger range of integers to work with.

6.1.2.3 The Real Data-Type

The real data-type is used to represent numbers that have a fractional part. For instance, we might wish to work with the real number –45.375. These types of numbers, because of their fractions, require a more complex representational strategy than integers. The method used to store these numbers is based on what we call the *scientific notation* representation of the number. In our example, $-45.375 = -4.5375 \times 10^1$, in scientific notation. When the number is written in scientific notation, we see that the number can be fully represented by three pieces of information: the sign of the number, the exponent, and the mantissa. The exponent is the power of 10 used in the notation, and the mantissa is the fraction and leading digit of the notation. (In our example, the sign is negative, the exponent is 1, and the mantissa would be 4.5375.)

FIGURE 6.1 Floating-point format.

A real number, once converted into scientific notation, can be represented in a computer word, by splitting the word up into three fields, as shown in Figure 6.1. This representation of a real value is called *floating-point format*, emphasizing that, when representing a number in scientific notation, the decimal point is moved, or floated, to a standard location.

The floating-point format is discussed further when we cover computer arithmetic, but for now, let us make one last observation about this notation. The floating-point representation is stored in a 32-bit computer word. It is limited by this word size to a certain precision. This precision turns out to be inadequate for many real calculations. Because of this, there are actually two sizes of floating-point representations: the one with the 32-bit word, and a larger representation using two words, or 64 bits. In many programming languages, like C++, you might be familiar with these two representations, that manifest themselves as the data-types *float*, which is the single precision 32-bit representation, and *double*, which is the double-precision 64-bit representation.

6.1.2.4 The Boolean Data-Type

Boolean data, as we already know, only consists of the two values, *true* and *false*. We already know that this data can be stored using one bit. However, our memory units, as described in Section 4.1, are set up to only access memories in bytes, or in full words. Storing and accessing Boolean values as one bit would cause problems. So instead, a Boolean value is typically represented as an 8-bit byte. By convention, if the byte is zero, the value being represented is considered to be the value *false*. Any other value (a non-zero value) is then considered to be the value *true*.

6.1.2.5 The Character Data-Type

On the computer, all data must be stored as binary numbers. When it comes to character data, each character must be represented as a binary number. For example, the character 'A' (capital A) must be stored as a number, and every other character on the keyboard must also be represented by a unique number.

All characters on the standard keyboard, and then some, can be represented with $256 = 2^8$ codes. The code size is then 8-bits. As mentioned in Section 4.1, the eight bits used to store character codes are referred to as a *byte*.

Assigning codes to individual characters has become standardized. The most widely used system is the ASCII (American Standard Code for Information Interchange) system.[2]

---

[2]ASCII was originally an improvement to an older code, EBCDIC (Extended Binary Coded Decimal Interchange Code), used on IBM machinery. ASCII code originally was a 7-bit code. The seven bits allowed representation of all the keys on the keyboard. Later ASCII was extended to 8 bits. The extra bit allowed the representation of other, mostly special characters.

Computers have become more international over the years. When ASCII was developed, most of the users of computers were English speaking. That is no longer the case. More and more users are now people who speak and write documents in languages other than English. Some of these users work in languages that do not use the Latin alphabet, as does English. For example, you might have a word processor document that is written in Russian, using the Cyrillic alphabet, or in Chinese, using Chinese characters. The problem is that the ASCII code only supports Latin characters from the standard QWERTY keyboard.

The internationalization of computing led to the development of UNICODE in 1987. This character code is a 16-bit code, developed from ASCII. As a matter of fact, the translation from ASCII to UNICODE is trivial; to form the UNICODE representation of an ASCII character, you simply fill the high-order byte, the prefix byte, with zeros, and copy the ASCII code for the character into the lower byte. So, for example, the ASCII code for 'A' is 01000001. The UNICODE representation is 00000000 01000001.

Codes for scripts other than the Latin script have prefix bytes that are not zero. UNICODE has enough room in its range, to accommodate all modern scripts, and then some.

6.1.2.6 Binary Coded Decimal

Some processors work with Binary Coded Decimal (BCD) data. BCD is a representation for integer data, just as is our normal positional binary representation.

The problem with using positional binary representations of numbers is that human users typically work in decimal. And so, you can expect that a human user would submit input to the computer in decimal, and expect output in decimal. If a processor works with a positional binary representation, it must then spend a great deal of time and effort converting from decimal to binary, and vice versa.

A solution to the conversion problem, is to build a processor that functions in decimal. But this is difficult, since most of our design work is built on Boolean algebra, which with its two value data, naturally leads to binary arithmetic. As a result, we probably would be more inclined to build a binary processor, but use a numeric representation that is easier to convert to and from decimal than our normal binary representation.

BCD represents a decimal number as a string of *nibbles*. A nibble is a 4-bit number. Each nibble is used to represent one of the digits in the decimal number. This strategy might best be demonstrated with an example. Suppose that you wish to represent the decimal number 365 in BCD. To do this, we start with the 4-bit representation of a 3, 0011. Follow this with the 4-bit representation of a 6, 0110, and then with the representation of a 5, 0101, to form the string 0011 0110 0101, as one large binary number.

You should notice that although BCD representations are, in fact, binary, they are relatively easy to convert to and from decimal. To convert from decimal, each digit is converted to its 4-bit binary expansion, and the results are strung together. To convert from BCD to decimal, the reverse process is applied. You might ask, if the conversion is so easy, why do not all processors use BCD to represent integers? The answer to this question is that arithmetic on BCD values is much more complex than arithmetic on numbers represented using a straight positional binary representation.

6.1.2.7 Data Manipulation Operation Types

As already stated, data manipulation instructions operate on data. We can categorize the types of operations they perform into a few categories.

- Arithmetic operations.

- Logic operations.

- Shift operations.

6.1.2.8 Arithmetic Operations

Arithmetic operations are operations that perform arithmetic on values stored in registers. As an example you might have a machine instruction that adds two registers together, as in the following.

```
add R0, R1
```

A typical processor would support, minimally, the usual addition operators, *add* and *subtract*, and possibly the multiplicative operators, *multiply* and *divide*.

Arithmetic must be supported for all numeric data-types that the processor implements. So for example, often you might see several instructions that perform addition. You might see the instruction

```
iadd R0, R1
```

to add two integers in the registers R0 and R1, and you might also have an instruction

```
fadd F0, F1
```

to add two floating-point numbers, stored in the floating-point registers F0 and F1.

6.1.2.9 Logic Operations

Logical operators perform bit-wise logical operations on registers. Bit-wise operation was explained in Section 5.2.3.2. Minimally a processor would support the standard three primitive operations of AND, OR, and NOT. So for instance your processor might implement the instruction

```
and R0, R1
```

that ANDs the contents of register R1 with register R0. Logical operations allow us to manipulate individual bits in a word. For instance, suppose that we wish to set Bit 4 of R0. We would do this with the following instruction.

$$\text{or R0, \#00010000b} \tag{6.1}$$

In Example 6.1 we assume that R0 is an 8-bit register. It is being ORed with a binary constant 00010000, which has all bits clear, except for Bit 4. When the constant, often called a logical *mask*, is ORed with R0, any bit in R0 at a position where the mask has a 0 bit, will remain the same. Any bit in R0 at a position where the mask has a 1 bit will be set to 1. In Example 6.1, only Bit 4 is 1 in the mask, and so the result of this operation is that all other bits in R0 will remain unchanged, and only Bit 4 will be set to a 1.

The AND operator is used to check the value of a bit, as opposed to setting the value of a bit. It allows you to rephrase the question of, say, whether Bit 4 is set to a question of whether a register contains zero. In the following code sequence we show how this works.

$$\text{and R0, \#00010000b} \tag{6.2}$$

The AND instruction will clear all positions in R0 where the mask has a 0. Only positions in R0 in which the mask has a 1 will remain unchanged. In Example 6.2, all bits in R0 are cleared, except Bit 4. So, if Bit 4 of R0 is initially 0, after the AND operation, R0 will contain a zero (all bits will be 0). If Bit 4 is initially a 1, after the AND operation, R0 will contain a non-zero value (all bits, except Bit 4, will be clear). It is now possible to check the result to see if it is zero, indicating that Bit 4 of R0 was originally 0, or if the result is non-zero, indicating that Bit 4 of R0 was originally 1.

6.1.2.10 Shift Operations

Shift operators perform shift operations on registers. Typically, a processor would at least support a logical shift, in both the right and left direction. As an example, your processor might have an instruction

```
shl R0
```

which would shift the contents of R0 left one bit. In addition, the processor might also support the circular and arithmetic shifts.

Shift operations are fairly important. Remember, from Section 5.2.3.3, that they are an easy way of implementing multiplication by powers of 2. By using them with addition, they can be used to do multiplication by any arbitrary small constant. For instance, suppose that you wish to multiply the contents of register R0 by 5. This can be accomplished with the following code sequence.

$$
\begin{array}{l}
\text{mov R1, R0} \\
\text{shl R0} \\
\text{shl R0} \\
\text{add R0, R1}
\end{array}
\qquad (6.3)
$$

In this sequence, the value of R0 is first saved in R1. Then R0 is multiplied by 4, by shifting it left by two bits. Finally, the previous value of R0 is added to R0, resulting in $R0 \times 4 + R0 = 5 \times R0$.

It is useful to compare this sequence with a straightforward multiply instruction

```
mult R0, #5
```

which would multiply R0 by the constant 5, using a multiplier circuit. The multiplier circuit is designed to be able to do any multiplication, and as we will see when we talk about computer arithmetic, because of its generality, it is slow. As it turns out, doing multiplication using the type of shift-add sequence we have presented is often faster than a full multiplication, even though the number of machine instructions is greater than the single multiplication instruction.

6.1.3 Control Operations

Control instructions are a bit different than data transfer or data manipulation instructions. Transfer and manipulation instructions operate on data stored in registers. Control instructions operate on the *flow of control* of the whole program. The *flow of control* is a phrase used to describe the sequence in which instructions are executed. By default, the processor uses *sequential control*, meaning that it executes instructions in sequence. For example, if the processor is executing a program in memory, and it is currently executing an instruction at Address 200, the next instruction to be executed will be the next instruction in memory, maybe at Address 201. Control instructions alter the flow of control, changing what instruction will be executed next, from the sequentially next instruction, to something else.

We can further categorize control instructions into sub-categories, which include most of the common types of control.

- Unconditional branches.

- Conditional branches.

- Machine reset.

- Context manipulation.

6.1.3.1 Unconditional Branches

An unconditional branch is an instruction that describes the next instruction to be executed by giving its address. This is demonstrated in the following code skeleton.

$$
\begin{array}{l}
\vdots \\
\texttt{jump xyz} \\
\vdots \\
\texttt{xyz:} \\
\vdots
\end{array}
\tag{6.4}
$$

In this skeleton the *jump* instruction is the unconditional branch. Its operand is a *label*, *xyz*, in assembly language. In machine language, this label represents an address, like Address 200. For this reason, labels are often referred to as *symbolic addresses*.

When the *jump* instruction is executed, the next instruction executed is the instruction at location *xyz*, rather than the next instruction after the *jump*. The *jump* instruction is called an unconditional branch because the jump is always performed. Under no circumstance would you not execute the jump, and execute the next instruction.

6.1.3.2 Conditional Branches

The conditional branch is taken under certain conditions, and ignored otherwise. Depending on the processor, there can be many different types of conditional branches. Most of them check results of arithmetic. For example, the processor might check to see if the contents of a register is zero, or non-zero. This is illustrated in the following code skeleton.

$$\vdots$$

```
                    beq R0, xyz
```

$$\vdots$$ (6.5)

```
xyz:
```

$$\vdots$$

The *beq* instruction (branch if equal) examines the value stored in the register R0. If the value stored is equal to zero, then the branch is taken. If not, execution continues with the instruction following the *beq* instruction, as normal.

There are two different common ways of implementing conditional branches. Example 6.5 uses a method called an *arithmetic conditional branch*. The name *arithmetic branch* refers to the fact that the instruction must do some work, some arithmetic, to determine if the branch should be taken. In an alternate scheme, called a *status-flag branch*, the conditional branch instruction does no arithmetic. The same example would be written for a status flag machine as follows.

$$\vdots$$

```
                    sub R0, #0
                    bz R0, xyz
```

$$\vdots$$ (6.6)

```
xyz:
```

$$\vdots$$

On a status-flag machine, there is a processor register, which might be called the FLAGS register. The FLAGS register is a collection of individual 1-bit status flags. When the CPU performs an arithmetic operation, it not only computes the result, but also sets the status-flags according to the result. In Example 6.6 we are presuming the existence of a status flag called Z. This status-flag indicates that the result of the last arithmetic operation was zero. In a normal processor several more status-flags would be available: maybe including a P flag, indicating a positive result, and an N flag, indicating a negative result.

In Example 6.6, a subtraction is performed, in which the constant 0 is subtracted from the register R0. This has no effect on R0, and is done solely to cause the processor to set, or clear, the Z status flag. If R0 contains 0, the subtraction will result in a value of 0, and the Z flag will be set. If R0 contains a non-zero value, the result of the subtraction will be non-zero, and the Z flag will be cleared. Once the Z flag has been given a correct value, the *bz* instruction uses it to determine whether or not to take the branch. If the *Z* flag is set, the branch is taken, and if the *Z* flag is cleared, execution continues sequentially.

6.1.3.3 Machine Reset Instructions

A machine reset instruction is used to turn the machine off. This type of instruction might also be called a *halt* instruction, and might appear as follows in assembly language.

```
halt
```

You can see that there isn't much to this instruction; it has no operands, and does a very simple job. The *halt* instruction, as mentioned, turns off the processor. The result of this is that the computer stops executing instructions. Visually, you might see the computer freeze, or reboot. This type of instruction is not something that most programs would include. Notice that it does not cause a program to stop, but rather causes the whole machine to stall out. Normally when a user wishes a program to terminate, the intent is not to have the machine freeze up, but rather to hand control to the operating system, which then kills the program, and reclaim its workspace in memory.

You might ask, who would use the *halt* instruction. The answer is that, as you probably have observed, there are programs that allow a user to power down the machine. These programs are low level programs in the operating system. If you were writing operating system code, you might very well be writing one of these low level drivers, and you would then use the *halt* instruction.

6.1.3.4 Context Manipulation Instructions

The last type of control instruction is the context manipulation instruction. These instruction are difficult to describe without first explaining what context switching is. Before discussing context, however, let us further classify context manipulation instructions into two different types.

- Subroutine instructions.
- Interrupt instructions.

These two types of instructions implement operations that are roughly the same, but are done for two different reasons.

Subroutine Instructions

Subroutine instructions are used to implement what are called procedures, or functions, in high-level languages. Consider the following high-level language skeleton.

$$
\begin{array}{l}
\vdots \\
\texttt{f ()} \\
\vdots \\
\texttt{function f ()} \qquad\qquad\qquad (6.7)\\
\texttt{begin} \\
\qquad \vdots \\
\qquad\qquad \texttt{return} \\
\texttt{end}
\end{array}
$$

The skeleton starts with a function call to the function f. As you should already know, when this statement is executed, control jumps down to the beginning of the function f, and the body of the function starts executing. Execution continues until the return statement at the end of the function is reached. Then control jumps back to the statement after the call to f, and execution continues in the main program.

Most of the mechanism used for a function call and return is just branching, which we have already discussed. The tricky part involves the return statement, which must jump to whatever location the function was called from. This location could be different on different calls to f. So, jump instructions are not adequate for this type of work, and we need special machine instructions to perform this operation. The use of these instructions is illustrated in the example below.

$$
\begin{array}{l}
\vdots \\
\qquad \texttt{call f} \\
\quad \vdots \\
\texttt{f :} \qquad\qquad\qquad\qquad (6.8)\\
\quad \vdots \\
\qquad \texttt{ret}
\end{array}
$$

This assembly skeleton behaves in the same manner as the high-level skeleton of Example 6.7. That is to say, the *call* instruction causes control to move to the address f. The code after location f is executed, until the *ret* instruction is executed, which sends control back to the instruction following the call instruction. The address of the instruction following the call is typically referred to as the *return address*, and a function or procedure in machine language is referred to as a *subroutine*.

Let us now tackle the question of how control finds its way back to the return address. The explanation is fairly straightforward. When the call instruction is executed, the return address is saved in a special place in memory. When the return instruction is executed, a jump is taken to the saved return address.

There is a complication to the above call-return scheme. The subroutine f might call another function g. In this case, while g is executing, the run-time system would have to maintain two saved return addresses: the return address for f, and the return address for g. What we have observed is that a single memory location, in which to save the return address for a call, will not work. We need several memory locations, one for each active call. This leads us to something called the *system stack*.

Most modern computers implement a system stack. The system stack is what would be called a LIFO (Last In First Out) data structure. Conceptually, a stack is a structure on top of which you can place objects. You only place, or *push*, objects on the top of the stack. You can also remove, or *pop*, an object off of the stack. Again, you are only allowed to remove the top object. So, we might say that a stack is a structure with two operations: *pop*, which removes the top element, and *push*, which adds a new top element.

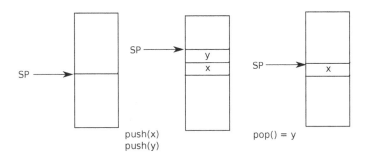

FIGURE 6.2 Illustrating the operations *pop* and *push*.

We illustrate the operations of a stack in Figure 6.2. The leftmost diagram shows a stack segment. The stack is implemented as a block of memory, and a top pointer is stored in an address register called SP (stack pointer). The bottom of the stack is at the bottom of the diagram. The top of the stack is above the bottom, at the position pointed to by the SP register. As the stack grows, the SP pointer decreases, moving up into lower memory.

In the middle diagram of Figure 6.2, two values have been pushed onto the stack: x and y, in that order. In the rightmost diagram, the stack has been popped. The top value, y, has been retrieved, and x is now the top element.

The stack example in Figure 6.2 should give us an understanding of how a stack can be used to save multiple values. We are now ready to describe how the call-return sequence differs from a simple jump. We start with the call instruction. A call instruction does the following.

1. Push the return address on to the system stack.

2. Jump to the subroutine address.

One question that surfaces, in the above call sequence, is how the processor knows the return address in Step 1. To answer this question, remember that the return address is the address of the next instruction after the call instruction. Well, when the call instruction is being executed, the address of the next instruction to be executed, which is the current value of the PC register, is, in fact the address of the instruction right after the call instruction. More simply stated, the return address is in the PC, and in Step 1 it suffices to push the value of the PC onto the system stack.

To clarify the workings of the call instruction, assume that a subroutine is being called. The call instruction is at address 400, the PC contains 401, and the subroutine is at address 200. In the first step, the call instruction would push 401 onto the stack. In the second step, the PC would be loaded with the value 200, causing a jump to that address.

The return instruction does the reverse of the call instruction.

1. Pop the return address off of the system stack.

2. Load the popped return address into the PC.

In the first step, the stored return address is removed from the stack, and in the second step, a jump to that address is performed by loading the return address into the PC.

Notice that this stack-based scheme works well for the described scenario in which the subroutine f calls a subroutine g. This is illustrated in Figure 6.3. First the main program calls f, causing its return address $(ra(f))$ to be pushed on the system stack. Next f calls g, causing the return address of g to be pushed. On return from g, the stack is popped back to the return address of f. Finally, on return from f, the stack is popped back to its original state, and control returns to the main program.

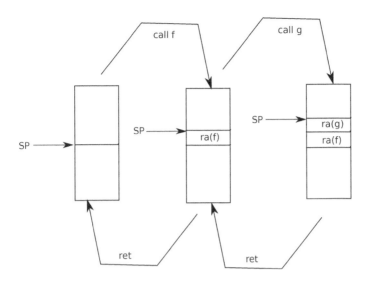

FIGURE 6.3 Nested subroutine activation example.

Interrupt Instructions

Interrupt instructions involve the interrupt mechanism described in Section 4.2.3. It was described in terms of I/O. Remember that when an I/O event occurs, control jumps from the executing program to an ISR, the code of the ISR is executed, and then control returns to the executing program. This should remind you of subroutines. The difference is that there is no call. The processor simply, in an uninvited manner, suspends the executing program and allows the ISR to take over control.

There are some other differences between subroutines and interrupts. Firstly, the interrupt mechanism, in general, does not use the stack. This makes sense when you realize that of the many different reasons for interrupting a program, some of them are to handle error conditions. In fact, an interrupt might be the indirect result of a corrupted stack. As a consequence, an ISR will typically store the return address in its own local memory. This local memory is used, not only to save the return address, but also any registers used in the ISR. Because the interrupted program must be able to resume after the interrupt, with no changes in its state, any register used by the program, and also used by the ISR, will need to be saved upon interrupt, and restored upon return. The storing and restoring of registers are often referred to as *context saving* and *context restoration*.

Notice that subroutines also store and restore context, just as do interrupt instructions, and for the same reason. Subroutines would store and restore context on the stack. Upon entry, a subroutine would push the registers it uses, and on return it would pop the values off the stack and back into the appropriate registers.

Although interrupts are usually forced on a program, it is actually possible for a program to request to be interrupted. This is illustrated with the following code skeleton.

$$\vdots$$

```
        syscall
```

$$\vdots$$

(6.9)

```
ISR:
```

$$\vdots$$

```
        iret
```

In this code, the program requests to be interrupted with the *syscall* instruction. This will cause a jump to the ISR. The ISR executes until it reaches the *iret* instruction, at which point it jumps back to the return address.

In terms of context saving, the *syscall* instruction mainly differs from the standard call instruction in that context is saved in the ISR, rather than on the stack, and no address is specified; the *syscall* jumps to a fixed address, namely the address of the ISR. Similarly, the *iret* instruction differs from the standard return instruction in that context is restored from the local memory of the ISR, rather than the stack.

You might wonder why a program would ask to be interrupted. The answer is that a user's program often operates under restrictions. For instance, a user's program would most likely not be allowed to manipulate an I/O device, like a printer, directly. The printer would, most likely, be shared with other users, and allowing a user program free hand with the printer would risk corrupting other users' jobs.

In fact, code operates in two different modes: *user mode* and *kernel mode*. In user mode, the code is restricted in terms of privileges. In kernel mode the code is allowed free hand to all resources. The problem then arises when the user code needs to use a resource that is protected from a user, like the printer. How can the user code be enabled to use the printer, with a degree of supervision? The answer is that the user program requests to be interrupted. The operating system then takes over. The operating system is always running in kernel mode. The user program passes information to the operating system, in a register, informing the operating system that it wishes to output to the printer. The operating system then serves as a proxy for the user program, taking the output from the program and sending it to the printer, using its kernel privileges.

So, the reason for the *syscall* instruction is so that a user program can request a *system service* from the operating system. There are many types of system services that can be requested. Some of them are concerned with I/O; others are not. For instance, one system service is often referred to as *exit to system*. When a program requests an exit to system, the operating system kills the program, and reclaims its workspace. Remember that when we talked about the halt instruction in Section 6.1.3.3, we observed that a program should not terminate with a *halt*, but rather do exactly what the exit to system service does. This then is the usual way a program terminates; with a *syscall* instruction.

6.2   INSTRUCTION FORMAT

We now turn to how to represent our instructions in machine language. We have been writing instructions in assembly language. We have done this to aid in comprehension, but as previously explained, our purpose is to describe machine language.

Machine language is a numeric language. So, we must describe how to translate an assembly language instruction, like the following, into a binary number.

$$\text{add R0, R1} \hspace{3cm} (6.10)$$

This instruction contains three pieces of information.

- The *operation*: the operation being performed. In this instruction it is addition.

- The *destination*: the operand where the result will be stored. In this instruction this would be R0.

- The *source*: the other operand. In this case this would be R1.

To translate Instruction 6.10 into a numeric representation, we might decide to store it in a single machine word. For demonstration purposes, let us assume that our word width is sixteen bits. What we will do is split this 16-bit word into three fields: one to store the operation, one to store the destination, and one to store the source. Each of these three pieces of information must still be converted into numbers, in order to make this scheme work.

FIGURE 6.4 Simple instruction format.

Figure 6.4 shows how the three fields might be arranged in the computer word. The operation consists of 6 bits, and both the destination and the source fields consist of 5 bits each. Each field would contain a binary number.

The operation field contains the *op-code* for the operation desired. Every operation the processor can perform is assigned a code, or number. Maybe subtraction is Operation 8, addition is Operation 9, etc. For our *add* instruction in Example 6.10, this field would then be 9.

FIGURE 6.5 Example machine instruction.

The destination, and source register can easily be converted into numeric form. We simply use the register number in the field. For Example 6.10, then, the destination would be 0, and the source would be 1. The full instruction is shown in Figure 6.5. The full machine instruction for our add example is the 16-bit number 0010010000000001.

There are a few pieces of information we can deduce from our instruction format. This information can be gleaned from the sizes of the three fields. We start with the size of the op-code, 6 bits. This field can specify numbers between 0 and $2^6 - 1 = 63$. What that tells us is that the processor has a total of at most 64 instructions in its ISA.

Moving on to the destination and source, we see that these two fields are 5 bits wide. Again, using the size of the fields, we know how many registers our processor has: $2^5 = 32$. And so, we see that there is a strict relationship between the sizes of the fields in a machine instruction, and the architecture of the processor.

The instruction format we have just discussed is proposed simply as a concept model. An actual instruction format is more complex, due to the inclusion of addressing modes, which we now need to explain.

6.3 ADDRESSING MODES

Our simple instruction format is a good start to understanding how machine languages are built. But, real machine languages tend to be more complex. One source of complexity is that, in our simple machine language, with the instruction format shown in Figure 6.4, only registers can be operands. This will not work, in general, if we wish to be able to use a memory unit. To use a memory unit, we need a mechanism that will allow us to draw operands from memory, sometimes, and from registers, other times. We need a mechanism that allows us to specify how to find an operand. This is what *addressing modes* do.

Consider the instruction format shown in Figure 6.4. This instruction contains two operand fields: the destination field, and the source field. Both of these fields contain numbers. Up until now we have stipulated that the numbers in these fields are interpreted as register numbers. So, for example, if the destination field contains the number 00111, we would expect to find the destination operand in Register 7. But when we start allowing operands from several devices, we may want this number to represent other things. For example, we might want this number to represent a memory address. Addressing modes are the specification of how the operand numbers are interpreted.

There are several common addressing modes, which we are about to list. Bear in mind that, most likely, not all of these addressing modes would be available on a particular processor. But, any processor will support at least a few of them.

Common addressing modes include the following.

- Direct mode.

- Indirect mode.

- Register direct mode.

- Register indirect mode.

- Immediate mode.

- Implicit mode

- Relative mode.

- Indexed mode.

We first demonstrate how these addressing modes work using assembly language, and later talk about how they are implemented in machine language.

6.3.1 Direct Mode

In direct mode addressing, the number given as an operand is a memory address. As an example, consider the following assembly instruction.

```
load R0, 5
```

In particular, look at the source operand. The operand given is 5. If this operand is interpreted in direct mode, the instruction specifies that memory location 5 should be copied into register R0. Written in RTL notation this would be

$$R0 \leftarrow M[5].$$

The mode is called *direct*, because the operand gives, directly, the location of the operand.

RAM

FIGURE 6.6 Example of indirect mode.

6.3.2 Indirect Mode

In indirect mode, we might see an instruction like the following.

```
load R0, (5)
```

The parentheses specify that the operand number is to be interpreted as a pointer to the effective operand. In RTL, this would appear as follows.

$$R0 \leftarrow M[M[5]]$$

Figure 6.6 shows the interpretation of indirect mode. Memory location 5 contains a 9, which is used as a pointer to the effective operand. At memory location 9 is the value 3, which would be loaded into the register R0. Indirect mode is so termed because one level of indirection is needed to find the effective operand, from the number given in the instruction operand field. That is to say, the effective operand cannot be found, directly, with one memory access, as is true in direct mode, but rather it is found with a second indirect access.

6.3.3 Register Direct Mode

In register direct mode the number given in the operand field of the instruction is interpreted as a register number. That register contains the effective operand. This is what we used in our discussion of instruction format in Example 6.10. Another example, in assembly language, follows.

```
mov R0, R5
```

This would have the following meaning.

$$R0 \leftarrow R5$$

6.3.4 Register Indirect Mode

Register indirect mode corresponds to indirect mode, in that indirection is used to locate the effective operand. With register indirect mode, the pointer is in a register, rather than in memory. In assembly language we might write an instruction with register indirect mode as follows.

```
load R0, @R5
```

The @ symbol specifies that the operand is in register indirect mode, and so, the instruction has the following meaning.

$$R0 \leftarrow M[R5]$$

6.3.5 Immediate Mode

For immediate mode, the number given in the operand field is the effective operand. It is not necessary to find the operand; it is immediately available from the instruction. The immediate mode allows the user to work with constants, as is illustrated in the following assembly instruction.

```
mov R0, #5
```

The # symbol specifies immediate mode for the second operand, and so this instruction has the following meaning.

$$R0 \leftarrow 5$$

6.3.6 Implicit Mode

Some instructions have implicit operands. That is to say, although an operand is being used, it is not mentioned in the instruction. To illustrate this, consider the following two instructions.

- `call 5`

- `syscall`

The similarity of these two instructions has already been discussed. Both jump to a subroutine, after saving the return address. The *call* instruction jumps to a subroutine at location 5 in memory. But the question is, then, where does the *syscall* instruction jump to? The answer is that it jumps to the fixed location of the ISR. The location of the ISR is then an operand to the *syscall* instruction. It does not need to be mentioned explicitly, because the jump location for the *syscall* instruction is always the same. The ISR address is an example of an operand given in implicit mode.

6.3.7 Relative Mode

With the relative addressing mode, also called PC relative mode, it is easier to explain how it works, as opposed to why it exists. As to how it works, consider the following assembly instruction.

```
load R0, $5
```

The dollar sign symbol indicates relative mode. This instruction moves the contents of a memory location into the R0 register. The address of the memory location is computed as PC + 5. In other words, in relative addressing, the address is computed by adding the contents of the PC to the number given as the operand. The meaning of the *load* instruction can be written in RTL as follows.

$$R0 \leftarrow M[PC + 5]$$

When computing the address of the effective operand, we refer to the number in the instruction as an *offset*. The value in the PC is referred to as the *base*. A relative address is decomposed into a base address, pointing to the general location of the operand, and an offset giving the precise location, relative to the base. As an analogy, suppose that we are describing the location where our car is parked. We could give the location as a pair of longitude and latitude values. This way of describing the location of the car would be what is called an *absolute* location specification. Instead, we might opt to give the location of the car by saying that it is on 3rd Avenue, in the block between 29th and 30th Streets, five houses from the corner of 29th. This is a relative description, giving the location of the car as an offset of five houses, from a base location of the corner of 3rd and 29th.

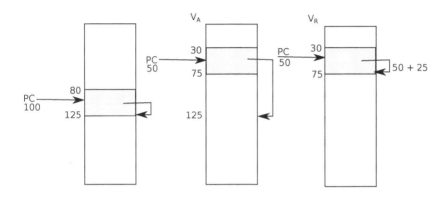

FIGURE 6.7 Code relocation example.

Turning to the question of why we would want to give an address in relative mode on a computer, consider a program that moves the contents of a memory location into the R0 register. The situation is shown in Figure 6.7. We can envision two versions of this program: one, V_A, uses the instruction

```
load R0, 125
```

which gives the address as an absolute address, and the other, V_R, uses the instruction

```
load R0, $25
```

giving the address relative to the PC register.

The diagram on the left of Figure 6.7 shows memory. The program, when running, is assigned to a workspace in mid-memory. The base address of the workspace is 80. When the move instruction is executed, the PC contains 100 as its value. The word being moved into R0 is at address 125.

The diagram in the middle shows what would happen if the program was relocated to a workspace lower in memory. Here, the workspace base address for V_A is 30. Because the operand address is absolute, Word 125 is still being moved into the R0 register. But this word is outside of the program's workspace. Programs are only allowed to access memory in their own workspace; not in the workspaces of other programs. As a consequence, V_A would most likely cause an access error when trying to do the load, and the program would be interrupted and aborted by the operating system.

The problem is that, using absolute addressing with relocation, the intent of the program is not taken into account. The move instruction in V_A will always use location 125 as the effective operand. The intent was to move the bottom word in the program workspace into R0. When the program is relocated to mid-memory, this word, coincidentally, has address 125. But, when relocated elsewhere, the address of this word must be given relative to the location of the program workspace.

The rightmost diagram in Figure 6.7 shows what happens if V_R is relocated in the same way as V_A. Because the workspace has been relocated to base location 30, when the move instruction is executed, the PC will contain 50, not 100. The offset 25 is added to it to calculate the address of the effective operand. This results in an address of 75, which is the intended location of the effective operand, at the bottom of the workspace. This version works because, as the program is relocated, the range of the PC is changed to coincide with the location of the workspace. Any address given relative to the PC is then also relative to the workspace.

6.3.8   Indexed Mode

The indexed addressing mode is very similar to relative mode, semantically, but it exists for a different reason. Indexed mode, also called *base addressing mode*, is illustrated by the following assembly language example.

```
load R0, 5(R1)
```

This instruction has the following meaning.

$$R0 \leftarrow M[R1 + 5]$$

In index mode, the address of the effective operand is given, not relative to the PC, but rather, relative to another *index register*. In this case, the index register is R1. Other than this difference, relative and indexed modes are the same; a constant offset is added to the contents of a base register to calculate the address of the effective operand.

The reason for the existence of index mode is to support what are often called *vector data structures*. A vector data structure is a data structure that consists of a collection of elements. Most programmers are familiar with the two most common types of vector data-types: arrays and records, or structures. In our discussion, we will focus on arrays.

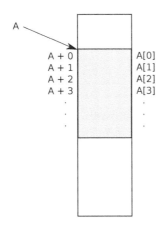

FIGURE 6.8 Array memory layout.

When you declare an array in a high-level language, at the machine level, the array is implemented as a block of memory. The base address of the array is the location in memory where the array starts. This word would be assigned to element $A[0]$. Consecutive elements in the array would then be assigned to consecutive words in memory. In other words, if the base address of the array A is called A, then $A[0]$ would be at $M[A]$, $A[1]$ would be at location $M[A+1]$, $A[2]$ would be at location $M[A + 2]$, etc. In general, $A[i]$ would be located at address $M[A + i]$. This layout is illustrated in Figure 6.8.

The indexed mode is used to support arrays in some high-level languages. Consider the high-level code below.

$$\text{for i = 0 to n-1 do} \\ \text{x = x + A[i]} \tag{6.11}$$

This loop sums the elements of the array A. We might ask how this loop would look in assembly language. The translation of Example 6.11 can be seen below.

```
        mov R1, #0                                          1
lab1:                                                       2
        load R0, n                                          3
        sub R0, #1                                          4
        sub R0, R1                                          5
        bz ext                                              6
        add x, A(R1)                                        7
        add R1, #1                                          8
        jump lab1                                           9
ext:                                                        10
```

Listing 6.1 Machine language translation for Example 6.15.

In Listing 6.1, i is stored in the register R1. So, we see that the first instruction in Line 1 initializes the variable i to 0. Then we enter the loop. The loop extends from Line 2 to Line 9. In the loop, the instructions in Lines 3 through 5 compute $n - 1 - i$, in the register R0. The bz instruction at Line 6, then branches to the loop exit label, ext, if the result is zero. The last two instructions in the loop, at Lines 7 and 8, add $A[i]$ to x, and increment the variable i.

Memory addresses in the code of Listing 6.1 are represented by symbolic addresses. So, we see the use of x, n, and A being used in place of numeric addresses. The address for $A[i]$ is given by A(R1), which uses indexed addressing mode. With our array layout in memory, $A[i]$ is computed by adding the value i to the base address of the array, A, resulting in $A + i$. This is done by the instruction on Line 7. (Remember that R1 is our variable i, from Example 6.11.) So, the indexed mode is designed to facilitate indexing of arrays, allowing us to use a register for the index variable, and supply the array base as the instruction operand.

6.3.9 Addressing in Machine Language

We now turn to the question of how addressing modes are implemented in machine language. This question can be restated as a question of how the assembly notation can be encoded as a number.

Returning to our example machine language of Section 6.2, remember that an instruction is stored in a 16-bit word. The op-code is 6 bits, and the two operand fields are 5 bits apiece. Let us now assume that the machine has 16 registers, and it also has a RAM unit of length 16 words. We are not going to try and implement all addressing modes discussed: only direct and register

op	DM	dst	SM	src
6	1	4	1	4

FIGURE 6.9 Instruction format with addressing modes.

direct. To implement these two addressing modes, we use an instruction format shown in Figure 6.9.

In our improved instruction format, the operand fields have been split in two. A 1-bit field is used to implement the addressing mode, and the actual operands are now represented by 4-bit fields. The destination field is split into *dst*, the operand, and DM (destination memory). The DM field indicates the addressing mode of the destination: 0, if the mode is register direct, and 1 if the addressing mode is direct. Likewise, the source field has been split into *src* and SM, with similar purposes, only for the source operand.

op	DM	dst	SM	src
001001	0	0000	1	1110

FIGURE 6.10 Machine code for Example 6.12.

To help digest this information, let us do an example. Consider the assembly instruction below.

$$\text{add R0, 14} \qquad (6.12)$$

This instruction uses register direct mode for the destination, and direct mode for the source. In machine language we would have the instruction shown in Figure 6.10. The DM field is 0 for register direct mode, and the SM field is 1 for direct mode.

To summarize how addressing mode is implemented, several bits are added for each operand field that indicate the mode. For our simple example only one bit was needed to choose between one of two modes. If the processor supports more modes, each mode would be given a multi-bit code, and we would need more bits per operand.

Notice that our RAM unit is very small. In most real computer systems, memory is relatively large. As a consequence, the number of bits to specify an address is much greater than the number of bits needed to specify a register number. For example, if the memory unit were just 64K long, 16 bits would be required to address memory, while only 4 bits would required to specify a register number. This bit difference causes complexity in instruction design. Although it is possible to make each instruction large enough to accommodate two 16-bit addresses, this wastes space, when register direct mode is used. A better solution, is to support variable-length instructions, allowing for a long instruction for use with direct mode, and a short instruction for use with register direct mode addressing. This option is discussed further in Section 6.5.3

6.4 ALTERNATE MACHINE ARCHITECTURES

We can categorize processors by their instruction format. Specifically we categorize machines by the number of operand fields in the machine instruction. So far, we have used instructions with two operands. But, this is not the only option; we might easily imagine a machine with more or fewer operands.

You might wonder why the number of operands in an instruction is an important categorization. There are several answers to this question. One is that the number of operands affects how high-level operations are performed on the machine. This brings up issues concerning execution efficiency. Another answer concerns the size of machine instructions. With more operands in an instruction, the larger the instruction. This brings up questions concerning space efficiency.

We will be talking about several different architectures. We list these below.

- Register machine (3-operand machine).

- Register implicit machine (2-operand machine).

- Accumulator machine (1-operand machine).

- Stack machine (0-operand machine).

Each architecture is known by a name, which is given first. Alternately, the machine might be described by how many operands each instruction contains. This description is given in parentheses.

For each of the architectures listed, we will be illustrating how the machine works, with a common, architecture-independent example. Our example machine, called the Reduced ISA Machine, or RIM machine, is a processor that has a collection of registers. The registers are numbered, 0 through 7. We refer to the registers in assembly language as R0, R1, R2, and all the way up to R7. More storage is available in a 256×16 memory unit. The word size is 16 bits.

As is normal, memory locations are accessed in assembly by specifying their addresses. With 256 words, all addresses are 8-bit. The addressing modes are direct, register direct, and immediate. The RIM machine also has a single I/O device for receiving input and transmitting output. All of the devices on the RIM machine are connected to a bus, using the bus architecture described in Chapter 4.

We now give a specification of the RIM machine's ISA. The RIM machine has the ability to perform the following operations.

- Data transfer.

  - Register-to-register.
  - Memory-to-register.
  - Register-to-memory.

- Arithmetic.

  - Addition.
  - Subtraction.

- Logic.

  - AND.
  - OR.
  - NOT.

- Control.

  - Jump.
  - Branch if zero.
  - Branch if not zero.

6.4.1 The Register Machine

We begin our discussion of alternate machine architectures with the register machine. A register machine, or 3-operand machine, has instructions with three operands. These operands could be called the destination, the first source operand, and the second source operand. Consider the following assembly instruction.

```
add R0, R1, R2
```

This is a typical 3-operand instruction. Its meaning is the following.

$$R0 \leftarrow R1 + R2$$

TABLE 6.1 ISA for the register machine.

Assembly Code	Machine Code	Meaning
load R_1, m	0000 R_1 m	$R_1 \leftarrow M[m]$
store m, γ	0001 γ m	$M[m] \leftarrow \gamma$
add R_1, γ_2, γ_3	0010 R_1 γ_2 γ_3	$R_1 \leftarrow \gamma_2 + \gamma_3, Z \leftarrow (\gamma_2 + \gamma_3) = 0$
sub R_1, γ_2, γ_3	0011 R_1 γ_2 γ_3	$R_1 \leftarrow \gamma_2 - \gamma_3, Z \leftarrow (\gamma_2 - \gamma_3) = 0$
and R_1, γ_2, γ_3	0100 R_1 γ_2 γ_3	$R_1 \leftarrow \gamma_2 \wedge \gamma_3, Z \leftarrow (\gamma_2 \wedge \gamma_3) = 0$
or R_1, γ_2, γ_3	0101 R_1 γ_2 γ_3	$R_1 \leftarrow \gamma_2 \vee \gamma_3, Z \leftarrow (\gamma_2 \vee \gamma_3) = 0$
not R_1, γ_2	0110 R_1 γ_2 0000	$R_1 \leftarrow \overline{\gamma_2}, Z \leftarrow \overline{\gamma_2} = 0$
jump m	0111 0000 m	$PC \leftarrow m$
bz m	1000 0000 m	if Z then $PC \leftarrow m$
bnz m	1001 0000 m	if $\overline{Z}$ then $PC \leftarrow m$

We see that the first operand given is the destination, the second is the first source, and the third operand is the second source.

6.4.1.1 Register Machine Instruction Format

We now turn our attention to the machine language for this machine. You may have already realized that the instruction format for this machine will have instructions with four major fields: an op-code, a destination, and two fields for the sources. In actuality, our machine will have two formats, which will be called the *register format* and the *memory format*. This is a common situation in processor design. We often find ourselves with instructions that are so semantically different that it is very difficult to represent all of them with just one format.

The instructions for our register machine are given in Table 6.1 This table has three columns. The first column gives the form of the instruction in assembly language. In the second column, the form of the instruction is shown in machine language, and in the third column the meaning of the instruction is given in RTL.

Let us examine a register format instruction from the table. The *add* instruction is a register format instruction. The assembly version of the instruction is the following.

add R_1, γ_2, γ_3

In this notation, we specify that the instruction takes three operands; the first operand is a register, and the other two are either registers or immediate values. (This is indicated by the notation γ.) Our machine has eight registers, numbered R0 to R7. The three operands can be any of these registers. In addition to register names, the source operands to the *add* instruction can be constants, written in the usual immediate mode notation, such as in the following instruction.

$$\text{add R0, R0, \#1} \tag{6.13}$$

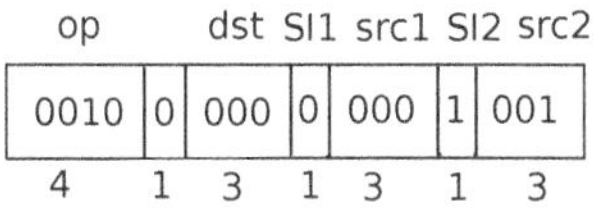

op   dst SI1 src1 SI2 src2

0010	0	000	0	000	1	001
4	1	3	1	3	1	3

FIGURE 6.11 Register instruction format.

To translate the assembly instruction into machine language we use the second column of Table 6.1. For the add instruction, the entry is

0010 R_1, γ_2, γ_3.

This notation is showing us the four fields of the register format. The register format is also illustrated in Figure 6.11. Figure 6.11 shows an instruction in a single word, with an op-code field of 4 bits, *op*, a destination field of 3 bits, *dst*, the first source operand field of 3 bits, *src1*, and the second source operand field of 3 bits, *src2*. In addition, each source operand has an addressing mode bit associated with it: *SI1* is for the first source, and *SI2* is for the second source. These bits are 0 if their corresponding operand is a register, and are 1 if their operand is an immediate value. There is an extra bit, right before the destination operand. This bit might be used as the addressing mode bit for the destination, except for the fact that the destination operand is always register direct. (You are not allowed to specify a constant as the destination of the operation.) And so, this bit is always 0.

Figure 6.11 also shows the *add* instruction from Example 6.13, translated into machine language. It starts with an op-code of 0010, indicating an addition, a destination of 000, indicating R0, a first source of 0 000, indicating R0, and finishes with the second source 1 001, indicating the constant #1.

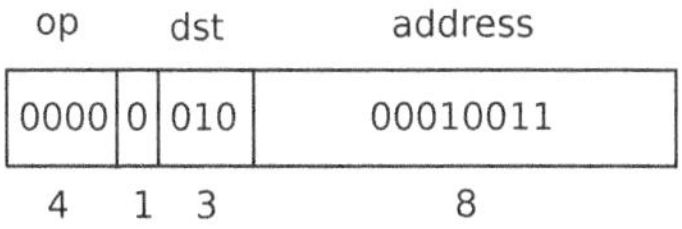

op   dst   address

0000	0	010	00010011
4	1	3	8

FIGURE 6.12 Memory instruction format.

The other instruction format, memory format, is illustrated by the *load* instruction. The assembly form is given by the following.

load R_1, m

Here the m represents a memory address. We see that this instruction is translated into machine languages as the following.

0000 R_1 m

The memory format is graphically illustrated in Figure 6.12. In memory format, the instruction is split into the op-code field, the destination field, and an address field. Figure 6.12 shows the values of these fields for the instruction below.

$$\text{load R2, \#19} \tag{6.14}$$

Although we call this machine a 3-operand machine, clearly the memory format instructions only have two operands. This is common, and still we would call the machine a 3-operand machine, because the data manipulation instructions are 3-operand.

The ISA for this machine implements a machine that is capable of doing all operations of our conceptual RIM machine. The *load* and *store* instructions allow us to transfer values between the processor and memory. The *add*, *sub*, *and*, *or*, and *not* instructions allow arithmetic, and logic operations, on values stored in registers. The *jump* instruction provides an unconditional branch, and the instructions *bz* and *bnz* perform conditional branches.

Conditional branches are performed on a status flag, as discussed in Section 6.1.3.2. For this machine we implement only the zero status flag, Z. Notice that all arithmetic and logic instructions set Z, and compute their results.

```
            add  R0,  #0 #0     ;  sum = 0                   1
            store  sum                                        2
            add  R0,  #0 #0     ;  i = 0                      3
            store  i                                          4
    lp :                        ;  while  i != x  do         5
            load  R0,  i                                      6
            load  R1,  x                                      7
            sub  R0,  R0,  R1                                 8
            bz  ext                                           9
            load  R0,  sum      ;  sum = sum + y             10
            load  R1,  y                                     11
```

```
        add  R0,  R0,  R1                              12
        store  sum,  R0                               13
        load  R0,  i        ;  i = i + 1              14
        add  R0,  R0,  #1                              15
        store  i,  R0                                 16
        jump  lp                                      17
ext:                                                  18
```

Listing 6.2 Register machine assembly multiplication of Example 6.15.

6.4.1.2 Register Machine Programming Example

Let us look at an example assembly language program. This program will multiply two numbers, stored in memory, and store the result in a third memory location. If the code were written in a higher-level language, it might resemble the following.

$$
\begin{aligned}
&\text{sum = 0} \\
&\text{i = 0} \\
&\text{while i != x do} \\
&\qquad \text{sum = sum + y} \\
&\qquad \text{i = i + 1}
\end{aligned}
\tag{6.15}
$$

This code adds the value y to the value sum, x times, thus leaving a result of $x \times y$ in sum. The assembly language version of this code is shown in Listing 6.2.

Listing 6.2 is augmented with comments (text following a semicolon). These comments show the high-level code implemented in the section following them. We start with Lines 1, and 2, which set the sum variable to 0. This is done by clearing R0 in Line 1, and then storing the result in sum. Lines 3 and 4 do a similar operation to clear the variable i. Notice that variable names, in the assembly code, are symbolic addresses for locations in memory.

At Line 5 we enter the while loop, and make the test to see if $i = x$. Lines 6 and 7 load the two variables into registers R0 and R1. The two registers are subtracted, in Line 8, which sets the Z status flag. In Line 9, a conditional branch sends control to outside the loop, if the result of the subtraction was zero.

TABLE 6.2   Machine code for Listing 6.2.

Address	Machine Code	Assembly code
00010100	0010 0 000 1 000 1 000	add R0, #0, #0
00010101	0001 0 000 00000000	store sum, R0
00010110	0010 0 000 1 000 1 000	add R0, #0, #0
00010111	0001 0 000 00000001	store i, R0
00011000	0000 0 000 00000001	load R0, i
00011001	0000 0 001 00000010	load R1, x
00011010	0011 0 000 0 000 0 001	sub R0, R0, R1
00011011	1000 0 000 00100100	bz ext
00011100	0000 0 000 00000000	load R0, sum
00011101	0000 0 001 00000011	load R1, y
00011110	0010 0 000 0 000 0 001	add R0, R0, R1
00011111	0001 0 000 00000000	store sum, R0
00100000	0000 0 000 00000001	load R0, i
00100001	0010 0 000 0 000 1 001	add R0, R0, #1
00100010	0001 0 000 00000001	store i, R0
00100011	0111 0 000 00011000	jump lp

The body of the *while* loop is composed of Lines 10 through 16. First, y is added to *sum*. Lines 10 and 11 load the two variables into the processor, Line 12 performs the addition, and Line 13 stores the result back to memory. Next, the index variable, i, is incremented. Line 14 loads i into the processor, Line 15 increments it, and Line 16 stores the result back to memory. Finally, Line 17 sends execution back up to the beginning of the loop.

Let us now consider how we might translate the assembly code in Listing 6.2 into machine language. To simplify this process, we will assume that our workspace is all of memory. As is typical, we split memory up into segments. For our purposes, we use two segments: a *data segment* in which we store all of our variables, and a *code segment*, in which we store the machine instructions. We will assume that the data segment will start at Address 0, and occupy memory up to Address 19. The code segment will then start at Address 20, and occupy the rest of memory. The important implications of this segmentation process are that the code starts at address 20, and the variables in our code will all have addresses less than 20.

To begin the translation, we need to agree on the memory locations for the variables. Let us use the following assignments: $sum \equiv M[0]$, $i \equiv M[1]$, $x \equiv M[2]$, and $y \equiv M[3]$. With these assignments, the machine code we would get would look like that in Table 6.2.

Assembling this code is a mechanical process. In the column for the address, we simply start counting in binary, starting from 20, with consecutive addresses on consecutive lines of the table. Assembly instructions are assembled using Table 6.1, and our variable address assignments. We show this with two examples.

To assemble the instruction

```
sub R0, R0, R1
```

we observe that the op-code for subtraction in Table 6.1 is 0001. We see that this is followed by a 1-bit 0, then the destination, as 3 bits, 000, for R0, then the address mode for the first source, 0, then the first source, as 3 bits, 000, then the addressing mode of the second source, 0, and finally the second source, 001, for R1. The result is shown in Table 6.2 at address 00011010.

As a second example, consider the instruction

```
bz ext
```

From Table 6.1 we see that the op-code for the bz instruction is 1000. This should be followed by 0 000, and then the target address of the jump, as eight bits, 00100100, which is one more than the address of the last instruction in the program. The final assembled instruction is shown in Table 6.2 at address 00011011.

6.4.2 The Register Implicit Machine

A register implicit machine is a machine in which the instructions have two operands. We eliminate one of the operands of the 3-operand machine by combining the destination operand with one of the source operands. That is to say that the destination operand of the 2-operand machine is actually one of the sources also. An example instruction for the register implicit machine might be the following.

```
add R0, R1
```

This instruction adds the contents of register R0 and register R1, and writes the result to R0.

6.4.2.1 Register Implicit Machine Instruction Format

We now look at the RIM machine, implemented as a register implicit machine. Figure 6.13 gives the instruction format for this machine. The top diagram shows that this version of the RIM machine will have only one instruction format. The format starts with a 4-bit op-code. This is followed by an addressing mode bit, M, the destination register as a 3-bit value, and an 8-bit

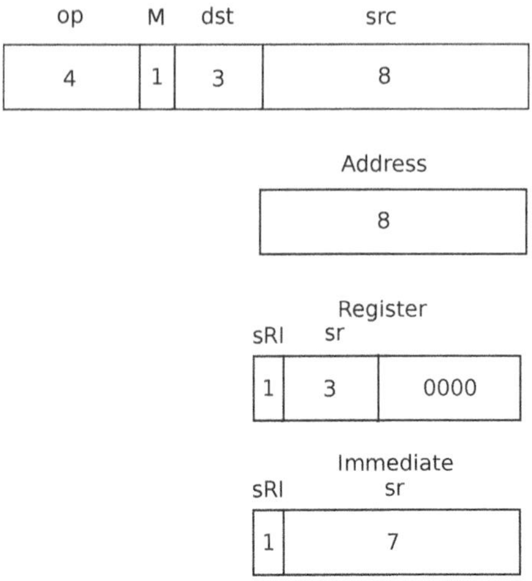

FIGURE 6.13 Register implicit instruction format.

value for the source operand. The source operand can be either a register direct specification, an immediate value, or a direct specification. These three options are shown in the middle diagram and the bottom diagram. For a direct specification, all eight bits of the source operand are used as an address. The direct option is indicated when the M (mode) bit is a 1. For the register direct and immediate modes, the high-order bit of the source operand, sRI, is used as an addressing mode bit. If sRI is a 0, the source operand is in register direct mode. In this case the source field, sr, is formed of the leftmost three remaining bits. If the sRI field is 1, the source operand is in immediate mode, and consists of the remaining seven bits of the source field.

The instruction set of our example machine is listed in Table 6.3. Table 6.3 has two entries for the move instruction. The first, in assembly language, is called the *load* instruction. It allows the user to transfer a value from memory to a register. For example, the user program might specify the following.

```
load R0, 128
```

The second form allows the user program to transfer a value between registers, as in the following.

```
mov R0, R1
```

Although these two instructions have different assembly language names, they both have the same op-code, 0000, which is the op-code for an instruction that

TABLE 6.3 ISA for the register implicit RIM machine.

Assembly Code	Machine Code	Meaning
load R_1, m	0000 1 R_1 m	$R_1 \leftarrow M[m]$
mov R_1, γ_2	0000 0 R_1 γ_2	$R_1 \leftarrow \gamma_2$
store m, R_1	0001 0 R_1 m	$M[m] \leftarrow R_1$
add R_1, γ_2	0010 0 R_1 γ_2	$R_1 \leftarrow R_1 + \gamma_2, Z \leftarrow (R_1 + \gamma_2) = 0$
sub $R_1, \gamma_2,$	0011 0 R_1 γ_2	$R_1 \leftarrow R_1 - \gamma_2, Z \leftarrow (R_1 - \gamma_2) = 0$
and R_1, γ_2	0100 0 R_1 γ_2	$R_1 \leftarrow R_1 \wedge \gamma_2, Z \leftarrow (R_1 \wedge \gamma_2) = 0$
or R_1, γ_2	0101 0 R_1 γ_2	$R_1 \leftarrow R_1 \vee \gamma_2, Z \leftarrow (R_1 \vee \gamma_2) = 0$
not R_1	0110 0 R_1 00000000	$R_1 \leftarrow \overline{R_1}, Z \leftarrow \overline{R_1} = 0$
jump m	0111 0 000 m	$PC \leftarrow m$
bz m	1000 0 000 m	if Z then $PC \leftarrow m$
bnz m	1001 0 000 m	if $\overline{Z}$ then $PC \leftarrow m$

moves a value into a destination register. We will refer to this single machine instruction as the *movR/M* (Move from Memory, or Register) instruction. There is also a store instruction, allowing the user to move values from registers into memory.

For data manipulation, the register implicit machine has all of the operations that the register machine had. The only difference is that all of the operations have been modified from a 3-operand form, to a 2-operand form. Control instructions for the register implicit machine are exactly the same in assembly language as for the register machine.

Table 6.3 is used to write machine language from assembly language. When using the table to translate from assembly instructions to machine instructions, we perform the same translation as for the register machine, except for the handling of the γ notation. Remembering that γ_2 can be either a register or an immediate value, in either case, γ_2 translates into an 8-bit field. If γ_2 is a register, then it is translated into an addressing mode bit, *sRI*, of 0, followed by a 3-bit register number, and then followed by four bits of zero. If γ_2 is a constant, it translates into an addressing mode bit of 1, followed by a 7-bit binary constant. We will illustrate this with several examples, using the move and load instructions. Consider the following instruction.

```
mov R0, R1
```

This instruction, in machine language, would start with the op-code 0000, continue with the M field of 0, then it would be followed by 000 for R0, then with 0 001 for register direct mode, and the operand R1, and finally a 0000 for the unused bits. This results in the instruction 0000 0 000 0 001 0000.

Now consider the following instruction.

```
mov R2, #4
```

This instruction starts with op-code 0000, an M bit of 0, and 010, for R2. This is followed by 1 0000100, for immediate mode, and the constant 4.

As a final example, consider the following instruction, which uses direct addressing mode.

```
load R1, 127
```

This instruction assembles to 0000 1 001 01111111. The M bit indicates direct mode, and the address field is completely consumed by the address 127.

```
            mov R0, #0        ; sum = 0              1
            store sum R0                             2
            mov R0, #0        ; i = 0                3
            store i, R0                              4
    lp:                       ; while i != x do      5
            load R0, i                               6
            load R1, x                               7
            sub R0, R1                               8
            bz ext                                   9
            load R0, sum      ; sum = sum + y       10
            load R1, y                              11
            add R0, R1                              12
            store sum, R0                           13
            load R0, i        ; i = i + 1           14
            add R0, #1                              15
            store i, R0                             16
            jump lp                                 17
    ext:                                            18
```

Listing 6.3 Register implicit machine assembly for Example 6.15.

6.4.2.2 Register Implicit Machine Programming Example

Our example high-level program, Example 6.15, would translate into the code shown in Listing 6.3. There are only a few differences between the code for this version, and the code for the register machine version in Listing 6.2; the data manipulation instructions have been reworked using two operands as opposed to three operands. As it turns out, our ability to work with three operands,

TABLE 6.4   Machine code for Listing 6.3.

Address	Machine Code	Assembly code
00010100	0000 0 000 1 0000000	mov R0, #0
00010101	0001 0 000 00000000	store sum, R0
00010110	0000 0 000 1 0000000	mov R0, #0
00010111	0001 0 000 00000001	store i, R0
00011000	0000 1 000 00000001	load R0, i
00011001	0000 1 001 00000010	load R1, x
00011010	0011 0 000 0 001 0000	sub R0, R1
00011011	1000 0 000 00100100	bz ext
00011100	0000 1 000 00000000	load R0, sum
00011101	0000 1 001 00000011	load R1, y
00011110	0010 0 000 0 001 0000	add R0, R1
00011111	0001 0 000 00000000	store sum, R0
00100000	0000 1 000 00000001	load R0, i
00100001	0010 0 000 1 0000001	add R0, #1
00100010	0001 0 000 00000001	store i, R0
00100011	0111 0 000 00011000	jump lp

for this particular example, was not fully utilized. We easily dropped down to two operands, with very little observable penalty.

As a final exercise, we assemble our assembly program into machine code. This machine code is shown in Table 6.4. We make the same assumptions on variable assignments and code segment base address as we did when assembling Listing 6.2.

6.4.3   The Accumulator Machine

The accumulator machine has instructions with only one operand. The question arises, if an instruction has only one operand, how is it that binary arithmetic operations, that require two operands, are performed? The answer to this question is that one operand is always implicit. That is to say, the user specifies the location of one source operand. The destination operand is always a special register called the *accumulator* (AC). With this architecture, it is then possible to specify addition with an instruction like the following.

```
add R2
```

This instruction adds the contents of R2 to the contents of the AC register, and leaves the result in the AC.

All operations use the AC register. So, the operation

```
load 128
```

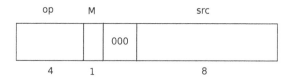

FIGURE 6.14 Accumulator instruction format.

TABLE 6.5 ISA for the RIM accumulator machine.

Assembly Code	Machine Code	Meaning
load m	0000 1 000 m	$AC \leftarrow M[m]$
load γ_2	0000 0 000 γ_2	$AC \leftarrow \gamma_2$
store m	0001 1 000 m	$M[m] \leftarrow AC$
store R_1	0001 0 000 0 R_1 0000	$R_1 \leftarrow AC$
add γ_2	0010 0 000 γ_2	$AC \leftarrow AC + \gamma_2,$ $Z \leftarrow (AC + \gamma_2) = 0$
sub $\gamma_2,$	0011 0 000 γ_2	$AC \leftarrow AC - \gamma_2,$ $Z \leftarrow (AC - \gamma_2) = 0$
and γ_2	0100 0 000 γ_2	$AC \leftarrow AC \wedge \gamma_2,$ $Z \leftarrow (AC \wedge R_2) = 0$
or γ_2	0101 0 000 γ_2	$AC \leftarrow AC \vee \gamma_2,$ $Z \leftarrow (AC \vee \gamma_2) = 0$
not	0110 0 000 00000000	$AC \leftarrow \overline{AC}, Z \leftarrow \overline{AC} = 0$
jump m	0111 0 000 m	$PC \leftarrow m$
bz m	1000 0 000 m	if Z then $PC \leftarrow m$
bnz m	1001 0 000 m	if $\overline{Z}$ then $PC \leftarrow m$

loads the value stored in $M[128]$ into the AC register. This uniformity of action results in a uniformity in instruction forms, allowing us to, just as with the register implicit machine, use only a single instruction format.

6.4.3.1 Accumulator Machine Instruction Format

The instruction format for the RIM accumulator machine is shown in Figure 6.14.Notice the striking similarity between the format shown in Figure 6.14, and that given for the register implicit machine in Figure 6.13. As a matter of fact, we are using exactly the same format. The only difference is that the destination field is now unused.

Table 6.5 gives the ISA for the accumulator machine. As already explained, the destination field is unused. Another trivial difference is that the move instruction has been renamed *load*. A more significant difference is that the store instruction is now capable of using a destination operand in both direct mode and register direct mode, allowing transfers from the accumulator to both memory and another register.

```
            load #0          ; sum = 0              1
            store sum                               2
            load #0          ; i = 0               3
            store i                                 4
    lp:                      ; while i != x do     5
            load x                                  6
            store R0                                7
            load i                                  8
            sub R0                                  9
            bz ext                                  10
            load y           ; sum = sum + y        11
            store R0                                12
            load sum                                13
            add R0                                  14
            store sum                               15
            load #1          ; i = i + 1            16
            store R0                                17
            load i                                  18
            add R0                                  19
            store i                                 20
            jump lp                                 21
    ext:                                            22
```

Listing 6.4 Accumulator machine assembly language for Example 6.15.

6.4.3.2 Accumulator Machine Programming Example

Our high-level program example, Example 6.15, could be written with assembly code similar to that in Listing 6.4. Notice how every value must pass through the AC in this code. For example, in Lines 1 and 2, to clear the variable sum, we first use the load instruction to clear the AC, and only then can we move the value from the AC to the variable sum, using a store instruction. In another example, when adding y to the variable sum, in Lines 11 through 15, first y is loaded into the AC, and only then can it be transferred into R0. Once transferred, this frees up the AC, so that sum can be loaded, and then the AC and R0 can be added.

TABLE 6.6 Accumulator machine code for Listing 6.4.

Address	Machine Code	Assembly code
00010100	0000 0 000 1 0000000	load #0
00010101	0001 1 000 00000000	store sum
00010110	0000 0 000 1 0000000	load #0
00010111	0001 1 000 00000001	store i
00011000	0000 1 000 00000010	load x
00011001	0001 0 000 0 000 0000	store R0
00011010	0000 1 000 00000001	load i
00011011	0011 0 000 0 000 0000	sub R0
00011100	1000 0 000 00101000	bz ext
00011101	0000 1 000 00000011	load y
00011110	0001 0 000 0 000 0000	store R0
00011111	0000 1 000 00000000	load sum
00100000	0010 0 000 0 000 0000	add R0
00100001	0001 1 000 00000000	store sum
00100010	0000 0 000 1 0000001	load #1
00100011	0001 0 000 0 000 0000	store R0
00100100	0000 1 000 00000001	load i
00100101	0010 0 000 0 000 0000	add R0
00100110	0001 1 000 00000001	store i
00100111	0111 0 000 00011000	jump lp

Table 6.6 shows the result of assembling the accumulator assembly code. The same assumptions on variable allocation apply as in the register machine.

6.4.4 The Stack Machine

The instructions for a stack machine have zero operands. This is an interesting statement. You might well ask how an operator can actually perform its operation without receiving any operands. You may already know the answer to this question, since we have discussed implicit mode operands. The answer is that the operands are located at implicit locations.

When working on a 1-operand machine, one of the operands is, implicitly, the AC register. The difference between a stack machine and an accumulator machine is that now all operands must be located automatically, without any information from the user program. Instead of using a single register, we now must use a collection of registers as implicit operands. The collection of registers used as implicit storage is called the *arithmetic stack*, and as the name implies, is a stack data structure. Remember that a stack is a structure supporting a *pop* operation and a *push* operation.

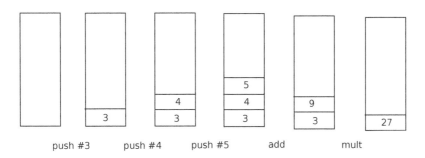

FIGURE 6.15 Stack arithmetic for Equation 6.17.

A stack can be used to manipulate data. The best way to illustrate this is with an example. Suppose that we wish to calculate

$$3 \times (4 + 5). \tag{6.16}$$

What we do is push the operands onto a stack. The operators are designed to automatically pop their operands off the stack, do their calculation, and push the result back onto the stack, so that it can be used by other operators. So, the following sequence of instructions might be used to do this calculation.

$$
\begin{array}{l}
\texttt{push \#3} \\
\texttt{push \#4} \\
\texttt{push \#5} \\
\texttt{add} \\
\texttt{mult}
\end{array}
\tag{6.17}
$$

The actions performed by this code sequence are illustrated in Figure 6.15. On the far left we see an empty stack. The **push #3** operation pushes a 3 onto the top of the stack in the next diagram. The next diagram shows the contents of the stack after the 4 is pushed. Then we show the stack after the 5 is pushed. The add instruction, like any operator, pops its two operands off the stack (the 5 and the 4), performs the addition operation, and pushes the result back onto the stack. In the last diagram, the multiplication operator does the same; it pops two operands off the stack, operates on them, and pushes the result.

The calculation in Equation 6.17 was relatively simple. We can perform much more elaborate calculations on the stack, interleaving pushes, pops, and operator applications.

Before we finish with Equation 6.17 and Figure 6.15, it is interesting to examine the format of our instructions. Remember that our machine is a 0-operand machine. While some of the instructions in Figure 6.15 are in fact

TABLE 6.7   ISA for the RIM stack machine.

Assembly Code	Machine Code	Meaning
push m	0000 0 000 m	$push(M[m])$
push i	0000 1 000 i	$push(i)$
pop m	0001 0 000 m	$M[m] \leftarrow pop$
add	0010 000000000000	$push(pop_1 + pop_2)$
sub	0011 000000000000	$push(pop_1 - pop_2)$
and	0100 000000000000	$push(pop_1 \wedge pop_2)$
or	0101 000000000000	$push(pop_1 \vee pop_2)$
not	0110 000000000000	$push(\overline{pop})$
jump	0111 000000000000	$PC \leftarrow pop$
bz	1000 000000000000	if $pop_1 = 0$ then $PC \leftarrow pop_2$
bnz	1001 000000000000	if $pop_1 \neq 0$ then $PC \leftarrow pop_2$

0-operand, specifically the *add* instruction and the *mult* instruction, all of the push instructions are, in fact, 1-operand instructions. This emphasizes the fact that it is very difficult to build a machine where no operand is ever specified. And, for this reason, 0-operand machines usually do have some instructions with operands. Even so, these machines are considered 0-operand machines, if their data manipulation instructions have no operands. In general, when categorizing a machine by number of operands, it is, in fact, the data manipulation instructions that determine the categorization.

6.4.4.1   Stack Machine Instruction Format

From Example 6.17, we see that we have two different types of instructions: the 1-operand instruction, and the 0-operand instruction. We can think of a 0-operand instruction as a 1-operand instruction that has an unused operand field. So, it is reasonable to design our machine with a single instruction format, namely a format for a 1-operand instruction. This format is shown in Figure 6.14. We are using the same format as we used for the accumulator machine. There are a couple of changes to the meaning of the fields, however. The M field will indicate direct mode, if 0, and immediate mode, if 1. Also, immediate values will be stored in the *src* field, using all 8 bits.

The ISA for our machine is given in Table 6.7. The instructions *pop* and *push* are used to transfer data to and from memory, respectively, from and to the stack. Notice that we do not have instructions for transferring data to and from registers. This is because the only registers we have, and need, are the registers that comprise the arithmetic stack. Notice, also, that we have a form of *push*, push i, for pushing an immediate mode value, i, onto the stack. All other instructions are the same as for the accumulator machine, except that they procure their operands off the stack.

When reading Table 6.7, the operation *push*, in the meaning column, indicates that the given value is pushed onto the top of the stack. Similarly, the operation *pop* indicates that a value is being popped off the top of the stack and used. The indexing on the pop operation is used when several values are popped in sequence. A subscript of 1 indicates the first value popped, and a subscript of 2 indicates the second value popped.

Notice our changes in branches. You will observe that the Z flag is no longer used. The conditional branches simply directly check the top element on the arithmetic stack to see if it is 0. If the top element is 0, that indicates that the last data manipulation operator produced a result of zero.

```
        push #0        ; sum = 0          1
        pop sum                           2
        push #0        ; i = 0            3
        pop i                             4
lp:                    ; while i != x do  5
        push #ext                         6
        push x                            7
        push i                            8
        sub                               9
        bz                                10
        push y         ; sum = sum + y   11
        push sum                         12
        add                              13
        pop sum                          14
        push #1        ; i = i + 1       15
        push i                           16
        add                              17
        pop i                            18
        push #lp                         19
        jump                             20
ext:                                     21
```

Listing 6.5 Stack machine assembly code for Example 6.15.

6.4.4.2 Stack Machine Programming Example

We are now ready to rewrite our multiplication example, Example 6.15, for the stack machine. The code is given in Listing 6.5. Most of the code should make sense. The sections that do the branches might be a bit confusing. For the *bz*

TABLE 6.8   Stack machine code for Listing 6.5.

Address	Machine Code	Assembly code
00010100	0000 1 000 00000000	push #0
00010101	0001 0 000 00000000	pop sum
00010110	0000 1 000 00000000	push #0
00010111	0001 0 000 00000001	pop i
00011000	0000 1 000 00100111	push #ext
00011001	0000 0 000 00000010	push x
00011010	0000 0 000 00000001	push i
00011011	0011 0 000 00000000	sub
00011100	1000 0 000 00000000	bz
00011101	0000 0 000 00000011	push y
00011110	0000 0 000 00000000	push sum
00011111	0010 0 000 00000000	add
00100000	0001 0 000 00000000	pop sum
00100001	0000 1 000 00000001	push #1
00100010	0000 0 000 00000001	push i
00100011	0010 0 000 00000000	add
00100100	0001 0 000 00000001	pop i
00100101	0000 1 000 00011000	push #lp
00100110	0111 0 000 00000000	jump

instruction, in Instructions 6 through 10, we first must push the address to be branched to. Then we push the two values being compared, and subtract them. At this point, the top element on the stack is the result of the subtraction, and immediately underneath it is the jump address. We then execute the bz instruction, which uses these two elements as its operands. In a similar fashion, for the jump instruction, in Instructions 19 and 20, we first must push the jump address, and then we execute the jump, which uses this element as its operand. Table 6.8 shows the machine code for the assembly code of Listing 6.5.

6.5   ISA DESIGN ISSUES

When discussing the ISA design, we have presented instruction sets, without going into the several choices made in the design process. Often, it might appear as if we are saying that there is only one rational way to design the instruction set. This could not be further from the truth. The truth is that at several points in time in design, you are faced with alternatives. You choose between them, base on your priorities. Quite often you are choosing between several trade-offs, and the solution you choose does not make you entirely happy, but it appears to be the best you can hope for.

In this section we examine some of the more basic considerations that you would deal with when designing an instruction set. These considerations are often at odds with each other.

6.5.1 Number of Registers

When you choose the number of registers in your processor, you are weighing several advantages and disadvantages against each other. Let's begin by assuming that your machine has hundreds of registers; more registers than you would know what to do with. This will help speed up your machine. This is because the CPU can access registers much faster than it can memory. Registers are inside the processor, so that if they are wisely used, the CPU does not have to wait for data to come in over the bus, and typically, registers do not have as much of the overhead of addressing, and can be accessed with less latency. With a large number of registers, data can be brought into the processor, and held there for use or reuse. If there are few registers, data must constantly be transferred in and out of memory, which, as mentioned, increases access latency.

But, there is also a disadvantage to having a lot of registers in the CPU. By adding numerous registers to the CPU, you are increasing the number of bits required for the op-code in an instruction, making it hard to design a compact instruction format with room for the op-code and all required operands. Also, the more register you include in the CPU, the larger the CPU becomes. Increasing the size of the circuitry usually results in slower circuitry. This slowing of the circuitry has to be balanced with the ability to hold data in the CPU. To balance these considerations, you typically try to keep the number of registers down, but still have enough so that you can bring in enough data for most common sequences of data manipulation instructions.

6.5.2 Word Size

There are several factors that influence word size. From a data viewpoint, you need to make the word wide enough to accommodate the data you are working with. However, if you make the word size too large for the data, many of the bits in the word will be unused when storing data, and a waste of space. Further, not only will the size of memory increase with the word width, the sizes of your registers and your data bus will also need to be increased.

Another consideration is the number of bits needed in your instructions. Particularly on a register machine, there is a lot of information that must be crammed into the instruction word: an op-code, a destination, and two source operands. So, while it is a good idea to keep your word size down, there are several forces at work that tend to result in a larger width.

6.5.3 Variable or Fixed-Length Instructions

With all of our RIM machines, we used fixed size instructions. That is to say, we decided that our word width was 16 bits, and that all instructions would fit into exactly one word. While this choice tends to simplify fetching an instruction from memory, it does introduce other problems. These problems become most pronounced with the stack and register machines.

The stack machine had one instruction format. It is shown in Figure 6.14, and includes fields for the op-code, the addressing mode, and the source operand. As you are already aware, the stack machine had two fundamentally different types of instructions: 1-operand instructions and 0-operand instructions. While the given fixed length format works well for the 1-operand instructions, it wastes excessive amounts of space, when used for the 0-operand instructions. So, for example, the *add* instruction contains the op-code 0010, and the rest of the bits of the instruction word are unused.

FIGURE 6.16 Eight-bit 0-operand format for the stack machine.

It might have been better to use two instruction formats, of different lengths. The 0-operand format would be 8 bits wide (a byte), and the 1-operand format would be the 16-bit format shown in Figure 6.14. A 0-operand format is shown in Figure 6.17. Although it wastes 4 bits, this is nothing compared to the twelve unused bits that would be wasted using the full word format.

The 3-operand register machine has the opposite problem of the stack machine. In the register machine, the concern is not wasting space, but rather not having enough space in the instruction for all three operands. In our register machine, the 16-bit register format works well with eight registers as operands. But, when used with immediate value operands, we must restrict the size of the immediate value to 3 bits, so that they can be fit into this format. It would be better, perhaps, to develop another 32-bit or larger format, for use when an immediate mode is specified, allowing full 16-bit immediate values.

Having several length instructions does, however, increase the complexity of fetching instructions from memory. With the single-word fixed-length instruction to fetch an instruction involved simply fetching the single word. With the variable length format, the byte containing the op-code must first be read, then, depending on the op-code value, the other bytes would be fetched, or would not be fetched. This decision process can increase circuit complexity, as well as increase the number of clock cycles required by the machine cycle.

6.5.4 Memory Access

When we design an instruction set we must decide which instructions are allowed to access memory. This issue may seem unimportant, but actually has a fairly sizable impact on machine speed.

It might, initially, seem that all instructions should be allowed to access memory. That is to say, any instruction can get operands from registers, or directly from memory. But on reflection we start realizing that memory access is significantly slower than register access. As a result, an instruction like an *add* instruction would be slowed down, significantly, if it were to fetch an operand from memory.

The issue becomes even more complex when we ask the question of how many clock cycles it takes to execute an *add* instruction. If we fix the number of cycles the *add* instruction consumes, then the number of cycles must be large enough to allow for a memory fetch. In cases where the operands are all fetched from registers, this fixed number of cycles would be excessive, and our processor would end up wasting a certain number of cycles.

The more efficient way of handling both memory and register fetches is to allow for a variable number of cycles for the *add* instruction. When register fetches are performed, only a small number of clock cycles would be consumed by the instruction. But, when a memory fetch is performed, the instruction would take a larger number of cycles. While this solution does address the problem of efficiency, its implementation increases the complexity of processor control.

Another consideration is that if we allow an instruction to access memory, we then have to find room in the instruction format for an address field, which often is fairly large. All of these considerations often lead to a design where only certain instructions can access memory. In our example RIM machines, we usually have only a few data transfer instructions that access memory. Data manipulation instructions only fetch from register. The rationale is that this will speed up the data manipulation instructions, and can speed up whole sections of code, if we are careful to fetch operands and hold them in registers until they are no longer needed, instead of fetching and re-fetching the operands each time they are used.

6.5.5 Instruction Set Size

This issue has been a huge topic of discussion in the recent past research literature. A trade-off exists between the advantages of a large set of complex instructions, and the advantages of a small set of simple instructions. Each of these viewpoints has its group of supporters.

6.5.5.1 RISC versus CISC Architectures

Machines that have large instruction sets and complex instructions are called Complex Instruction Set Computers (CISC machines). A complex instruction is an instruction that, rather than perform a single simple operation, performs several simple operations. For example, an instruction that loads a value from memory using indexed mode, and then automatically increments the index register would be considered complex. Normally this operation would be performed by two separate instructions: one would access memory, and the second would increment the index register. The reason this particular complex instruction might be included in an instruction set is because it is extremely useful when processing the elements of an array, sequentially. Including this instruction tends to make the computer easier to use, from the perspective of a programmer.

Machines that have small instruction sets, and only simple instructions, are classified as Reduced Instruction Set Computers (RISC machines).[3] When you limit the size of the instruction set, it simplifies the circuitry of the processor, and as a consequence tends to result in a faster processor. When you limit the complexity of instructions, you potentially can decrease the length of the machine cycle, further speeding up the machine. But, an equally important feature of the RISC architecture is that by decreasing the complexity of the machine, you make it much easier to use advanced architectural techniques, like *pipelining* (discussed in Chapter 10). Pipelining is a technique that allows the processor to work on several instructions at the same time, further enhancing the efficiency of the CPU.

You would think that the trade-off is a simple one, between the speed of a RISC machine, and the flexibility of a CISC machine. But, the situation is more complex than that. Although instructions on a RISC machine are faster, it takes more instructions to accomplish the same work that can be done on a CISC machine with fewer instructions. This would tend to slow down the RISC machine. Nonetheless, the idea that RISC architectures produce shorter machine cycles is worth keeping in mind, when designing an instruction set. It is also worth remembering that a few complex instructions in the set make

---

[3]RISC machines are characterized by several other properties. One is that RISC machines usually have heavy restrictions on memory access; usually only a few instructions are included in the instruction set to do memory fetches.

the machine easier to use. The result is often a compromise between the two extremes; an instruction set with some complexity, and yet with relatively small size.

6.5.5.2 Orthogonality and Completeness

The trade-off between RISC and CISC architectures can be rephrased. The terms *orthogonality* and *completeness* are often used to describe this trade-off. Both of these properties have advantages and disadvantages.

An instruction set is call orthogonal if any operation that the processor can perform is implemented by only one instruction. Probably the best way to illustrate orthogonality is to present a case that demonstrates the lack of orthogonality. Consider a processor that has an *add* instruction, as well as an *inc* instruction, that increments a register. The increment instruction might be used as follows.

```
inc R2
```

This same operation can also be performed using the *add* instruction, as follows.

```
add R2 #1
```

These two instructions are a demonstration that the instruction set for this processor is not totally orthogonal. The same operation can be performed by two different instruction sequences.

Orthogonality works to keep the size of the instruction set down. As discussed in Section 6.5.5.1, this tends to increase the speed of the processor. The problem with orthogonality is that it decreases the flexibility of the machine. So, in our example, the increment instruction is a genuinely convenient addition to the instruction set, from the user's perspective.

An instruction set is complete if every operation that the user of the machine requires is implemented with an instruction. So, for instance, if the user needs to multiply floating-point numbers, there is an instruction for doing this. If the user needs an instruction that branches if an arithmetic result is negative, this instruction is in the instruction set.

Obviously, completeness works to increase the flexibility of an instruction set. However, it also tends to increase the size of the instruction set. As with orthogonality, we see both advantages and disadvantages to completeness.

TABLE 6.9 I/O instruction definitions.

Assembly Code	Machine Code	Meaning
in R	1010 0 R 00000000	$R \leftarrow D_{in}$
out γ	1011 0 000 γ	$D_{out} \leftarrow \gamma$

As can be seen, this issue of instruction set size is fairly complex. It is also an important issue in processor design. It is quite understandable that it has been intensely scrutinized by the research community.

6.5.6 ISA

We have looked at several instruction set architectures: the register machine, the register implicit machine, the accumulator machine, and the stack machine. Of course, to design a processor, you must decide which of these architectures to use.

You may have noticed that as we move through the architectures, to fewer operands, we introduce more and more implicit mode operands. These are operands located in a fixed and standard location. This tends to simplify the processor wiring; we do not need circuitry to choose between several possible locations. This tends to speed up the execution of each instruction. On the other hand, as we reduce the number of operands, we need more data transfer instructions to move operands into and from implicit registers, where they can be operated on. This tends to increase the program size. This trade-off is very similar to the trade-off between RISC and CISC machines. Simplicity produces faster instructions, but increases the number of instructions required to do a job. And, again, this is an area where a well-thought-out compromise can result in a better machine.

6.6 THE BRIM MACHINE

Of the several architectures we considered, we will be using one in the chapters to come. The choice of our design model is fairly arbitrary, although we chose an architecture which is a compromise between the operand-heavy register machine, and the operand-scarce stack machine. Our design model is the register implicit machine. We will be using the model presented in this chapter, enhanced with special instruction I/O. We refer to this enhanced machine as the BRIM machine (Basic Register Implicit Machine).[4]

---

[4]An assembler for the BRIM machine is available as a distribution package from the publisher. The distribution contains the assembler, example programs, and further documentation on the assembler.

The BRIM ISA is specified in Table 6.3. To the register implicit machine we add an I/O device, allowing the user to provide input to the BRIM machine, and view the output of the machine. We add the two I/O instructions to the ISA of the RIM machine, with ISA modifications shown in Table 6.9. The *in* instruction moves the current value of the I/O device, into the destination register. The *out* instruction moves the value of the source register, or an immediate value, into the I/O device's device register.

6.7 SUMMARY

A great deal more can be said about computer architecture. We have tried to cover only the basics. Our focus was on the machine level, and machine language.

At the beginning of the chapter, we covered how to represent an assembly language instruction numerically. This involved storing the pertinent information in fields. Addressing modes were discussed. These modes allow the instruction to indicate from where operands are fetched.

Throughout our discussion, we worked with a simple conceptual machine. This machine demonstrated an architecture with a fixed size instruction. Using this abstract machine, we worked through several incarnations of the machine, demonstrating different instruction set architectures. We closed the chapter with a discussion of some of the more important instruction set design issues.

6.8 EXERCISES

6.1 For each of the following assembly instructions, give the contents of the register R1, after the instruction is executed. You may assume that $R3 = 9$, $PC = 11$, and that all memory locations contain one more than their address $(M[k] = k + 1)$.

    a. `load R1, 3`

    b. `mov R1, R3`

    c. `mov R1, #3`

    d. `load R1, @R3`

    e. `load R1, (3)`

    f. `load R1, $3`

    g. `load R1, 5(R3)`

6.2 Repeat Exercise 6.1 with $R3 = 7$, $PC = 9$, and $M[k] = 2k$.

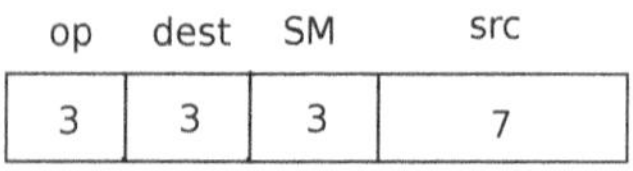

op	dest	SM	src
3	3	3	7

op   dest   SM   src

FIGURE 6.17   Move instruction format for Exercise 6.3.

6.3   Assume that the machine in Exercise 6.1 has a move instruction with the format shown in Figure 6.17. In this figure, the format has a standard 2-operand format. The destination operand is always in register direct mode. The source operand can be in any of the modes shown in Exercise 6.1. The SM field identifies the mode of the source operand. Assume that the op-code for the move instruction is 6, the op-code for the load instruction is 7, and that the addressing modes used in Exercise 6.1 are numbered, in order: 0 for direct mode, 1 for register direct mode, 2 for immediate mode, 3 for register indirect mode, 4 for indirect mode, 5 for relative mode, and 6 for index mode. For each of the following assembly language instructions, write the equivalent machine language instruction. (Note: for the index mode you will need to store both the offset and the index register number in the source field.)

a. mov R3, R2

b. load R0, 3

c. load R2, 4(R2)

d. load R0, (5)

e. load R1, $6

f. load R3, @R1

g. mov R0, #7

6.4   Translate the following high-level code, that calculates 2^n, into assembly language for the register machine of Section 6.4.1.

$$\begin{array}{l} \text{pr = 1} \\ \text{while n != 0 do} \\ \quad \text{pr = pr + pr} \\ \quad \text{n = n - 1} \end{array} \qquad \text{(P6.1)}$$

6.5   Translate the code segment from Exercise 6.4 into assembly language for the register implicit machine of Section 6.4.2.

6.6   Translate the code segment from Exercise 6.4 into assembly language for the accumulator machine of Section 6.4.3.

6.7 Translate the code segment from Exercise 6.4 into assembly language for the stack machine of Section 6.4.4.

6.8 Translate your assembly code from Exercise 6.4 into machine language. Assume that the variables are allocated as $pr = M[0]$, and $n = M[1]$, and that the code is loaded starting at $M[10]$.

6.9 Repeat Exercise 6.8, only using the assembly code from Exercise 6.5.

6.10 Repeat Exercise 6.8, only using the assembly code from Exercise 6.6.

6.11 Repeat Exercise 6.8, only using the assembly code from Exercise 6.7.

6.12 Design variable length instruction formats for the register implicit machine of Section 6.4.2. In your diagram, be specific; show field widths. Ensure that your op-code is one byte, and that your instruction lengths are always a whole number of bytes. Indicate which of the machine instructions from Table 6.3 use which formats, and if needed, explain what the fields of your formats are used for.

6.13 Write a BRIM program to input a value n, and output the sum $1 + 2 + 3 + \ldots + n = \sum_{i=1}^{n} i$.

6.14 Write a BRIM program to input a value n, and output n^2.

6.15 Write a BRIM program to input values n and m, and output $\lfloor \frac{n}{m} \rfloor$. (Note: you may need to determine if a value is greater than, or equal to zero, to do this exercise. Remember that a value is greater than, or equal to zero, if its sign bit is a 0.)

Hardwired CPU Design

CONTENTS

THE SUBJECT OF COMPUTER ORGANIZATION is a little strange to study. A great deal of effort must be expended to first learn all of the preliminary material, before you can actually start explaining the core material. In this book, up to this point, most of what we have covered can be considered as preliminary. We are now ready to start discussing how a computer is organized.

The goal of the discussion in this chapter is to build a functioning processor. To build a processor, we must choose which of several architectures we will be implementing. In Chapter 6 we described a basic conceptual machine, called the RIM machine. We implemented this machine as a register implicit machine, called the BRIM machine. In this chapter we continue with this work, by implementing the BRIM machine.

We will attempt to construct the BRIM machine with sufficient generality, so that the basic construction of the machine can be reused to implement a full register machine, an accumulator machine, and a stack machine. At the end of the chapter we will then explain how these machines might be constructed.

The processor is a sequential circuit. We specify it with an RTL description, and just as with any RTL implementation, this means that the resulting circuit is composed of a data-path, and a control circuit, that manipulate the data-path. The most challenging part of processor design is probably the construction of the control circuitry. The processor control circuit, as in RTL design, is the circuitry that manipulates a data-path. The design of the control circuit usually follows one of two schemes.

- Hardwire control.

- Micro-programmed control.

In the hardwired design, we design a control circuit much in the same way as we did when designing RTL control circuits, and the result is a piece of hardware. In micro-programmed design, the control circuit is designed as a processor that executes micro-instructions. In this scheme, the processor contains another smaller processor that executes the machine cycle, stored as a micro-program in some sort of firmware memory.

It is difficult to describe micro-programmed control, at this point, in any better detail. It becomes much easier to describe it after we have examined hardwired control. So, we start our discussion on processor design by describing hardwired control.

7.1   REGISTER IMPLICIT MACHINE DESIGN

As mentioned, in hardwired design we are essentially designing the processor using the same techniques we used when working with RTL instructions. Specifically, when we talked about RTL design in Section 5.3, we discussed how to design an algorithmic machine. These were machines that implement an algorithm in circuitry.

Processors are algorithmic machines; they execute an algorithm continuously. The algorithm, to which we refer, is the machine cycle. Algorithmic design uses a sequencer to implement timing. In that, and all other respects, the design process for the processor is standard for RTL machines. We first would describe a data-path, and then specify how the control circuit manipulates the data-path.

7.1.1 The Data-Path

In Chapter 4 we presented an architecture based on a bus connection, indicating that most computer systems use some sort of variation on this architecture. In Chapter 5 we built circuits from RTL descriptions. All of these circuits used a direct connect type of data-path. When we build the processor, however, we will be using the more usual bus connection for the data-path.

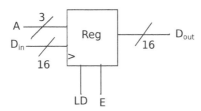

FIGURE 7.1 Register file interface.

7.1.1.1 Bus-Based Data-Path

Let us begin by describing the constraints on our data path. The BRIM machine, with its underlying RIM model, is more fully described in Chapter 6. We see that it has eight registers, numbered 0 through 7. We can think of these registers as a single device; a sort of memory unit, typically called a *register file*. The register file is very much like a clocked SRAM unit. The block diagram of the register file for the BRIM machine is shown in Figure 7.1. The 3-bit address port allows register numbers between 0 and 7, and the 16-bit data ports accommodate the 16-bit word width of the BRIM machine.

The BRIM machine also has a memory unit of size 256 × 16. This will be implemented using another standard SRAM unit. We will be implementing a full BRIM implementation of the RIM machine, and so the machine also has an I/O device.

We must be careful, when describing the data-path, to ensure that all of the instructions in the instruction set can be realized on the resulting circuitry. The BRIM machine has data transfer instructions. These will be implemented using the bus connection to move data between registers. The BRIM machine also has data manipulation instructions. These will require circuitry to perform the required operations. We collect this circuitry into a single device, called the Arithmetic Logic Shift Unit (ALSU, or ALU for short). Finally, the BRIM machine has control instructions. These will use the PC processor register. The subject of the Program Counter register brings up an important point. We will need several processor registers, separate from the general-purpose registers in the register file, to help the processor through the machine cycle.

The data-path for the RIM machine is shown in Figure 7.2. This figure is rather complex, so that it will take a bit of explanation to glean all the information available from it. We start by saying that this is a bus diagram, very similar to diagrams we presented in Chapter 4, such as Figure 4.2. The u-shaped cable which encloses the bus devices is actually the data bus. But then the question is, what has happened to the address bus, and the control bus?

Recall from Chapter 4 what the roles of the control and address buses were. The control bus was used to broadcast the operation required by the CPU to all bus devices. The address bus was, then, needed so that each device could determine whether it was the target of the request. In the scheme presented, we only needed a single write line in the control bus. But what if we decide to include multiple write lines in the bus, one for each device? This situation is illustrated in Figure 7.3. The figure shows two bus devices, D_0 and D_1. In the top diagram the devices are connected to a standard bus, with a bus address of 0 for D_0, and a bus address of 1 for D_1. A single control line, Wt, is used to request a write operation on one of the two devices. The bottom figure shows the same two devices, with a different bus structure. In this diagram, we have eliminated the address bus. Instead, we have two write control lines. Wt_0 is a dedicated line for D_0, and Wt_1 is dedicated to D_1. With this structure, when the CPU requires a write operation of D_1, it simply triggers the Wt_1 control line. No addressing is required.

The same type of structuring as in Figure 7.3 has been applied to the data-path of Figure 7.2. The data bus is shown, and the control bus is composed of all of the dedicated control lines in the diagram (lines like RLD, IRX, and PCIN). The address bus has been eliminated altogether, using a direct connection scheme, described momentarily, that supplies the internal address to the memory unit.

Another general observation about Figure 7.2 is that it contains a mix of devices: some of them are part of the CPU, and some are external to the CPU. The RAM unit and the I/O device would normally be external to the CPU. Every other device shown in the diagram would be internal to the CPU. Normally we think of the bus as being controlled by the CPU. But with the mix of internal and external devices in the bus diagram, we cannot treat the CPU as an external agent. Instead, we will be referring to the device controlling the bus as the *Control Unit* (CU). The CU is very much like the control circuitry

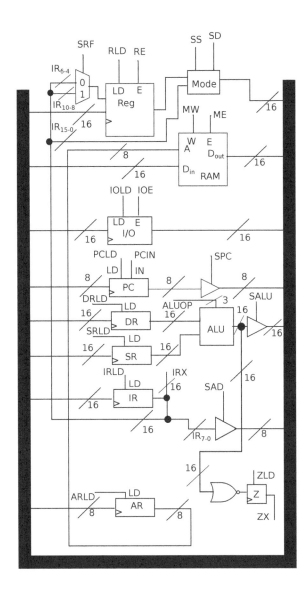

FIGURE 7.2 BRIM data-path.

that we designed for RTL circuits, and is one of the more important devices in the CPU.

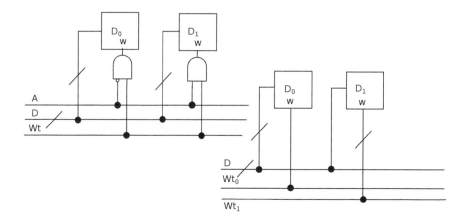

FIGURE 7.3 Eliminating the address bus.

Let us now examine the bus connections in Figure 7.2. In a bus architecture, devices are connected to the bus using the configuration shown in Figure 7.4. A tri-state switch is used to block the device output from flowing onto the bus. This switch is controlled by an enable control line. Input to the device is blocked using some sort of load control line. With standard registers, remember, the register might also have two other control lines: an increment line and a clear line.

Notice that some of the bus connections are 8-bit, and some are 16-bit. Let us clarify this situation. The data bus is 16 bits wide. Eight-bit connections are used to transfer memory addresses over the bus. When an 8-bit connection is made, only the lower 8 bits of the bus are connected; the upper bits are not used.

Taking an even more detailed look at Figure 7.2, we see that it contains a register file, a RAM unit, an I/O device, and six processor registers. These registers are as follows.

- The Program Counter (PC). This register, as previously discussed, contains the address of the next instruction to be executed. Addresses on the RIM machine are 8-bit, and so this register is 8 bits wide.

- The Destination Register (DR). This register is used to store the destination operand from the current instruction. It contains data, which, on the RIM machine, is 16 bits, and so this register is 16 bits wide.

- The Source Register (SR). This register is used to store the source

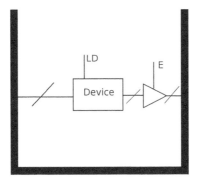

FIGURE 7.4 Standard bus connection.

operand from the current instruction. It is a data register, and so is 16 bits wide.

- The Instruction Register (IR). This register is used to store the current instruction, after it has been fetched from memory. Our instructions are one word wide, and so this register needs to be a 16-bit register.

- The Address Register (AR). This register is used to address the RAM unit. The RAM unit is hardwired to procure its address from this register. Since the AR stores addresses, it must be 8 bits wide.

- The Zero status flag (Z). This is a 1-bit register that functions as a zero status flag.

Once we understand the function of the processor registers in Figure 7.2, we then can look at the connections shown. These connections were drawn in accordance with the machine instructions for the BRIM machine, in Tables 6.3 and 6.9. We start with the move instruction. The *mov* instruction transfers data from somewhere else, into a register. All such transfers will be done via the bus. To move something into the destination register, we need connections to the input ports of the register file. The data will come over the data bus, and so the input data port of the register file is connected to the bus. The address port needs to be supplied with the register number for the destination register. This is done using the MUX connected to the address port. Option 1 for the MUX comes from the IR, bits 8 through 10. This is, in fact, the destination field of the current instruction.

The source operand for a move instruction can be given as a register number, a memory address, or an immediate value. It must be possible to place each of these three types of operands on the bus. We start with the register-direct operand. Option 0 of the address MUX allows the source register number, Bits 4 through 6 of the IR, to specify the address for the register file. In addition, there is a connection from the output data port of the register file, through the *mode multiplexer*, which is used to place a register-direct operand on the bus. Next we consider the direct operand, coming from the RAM unit. Again, there is a connection from the output data port of the RAM to the bus, allowing this type of operand to be handled. Finally, for the immediate value, we need a way of placing Bits 6 through 0 of the instruction on the bus. To do this, we have a connection from the IR to the mode MUX, and on to the bus.

Moving on to the store instruction, this instruction moves data from a register to a memory location. The register number is given in the instruction as the destination field, and the memory address is given as the address field. The RAM unit is addressed using the AR register, and so we must have a way of moving the contents of the instruction address field into the AR register. This is done by using the connection from the IR to the bus. This connection takes the rightmost 8 bits of the instruction, the address field, and puts them on the bus. From the bus, these bits can be loaded into the AR register, using its input connection to the bus.

For the store instruction, we must also transfer the contents of the specified register to the bus, where it can be loaded into the data-in connection of the RAM unit. This is done by moving Bits 10 through 8, the destination field of the IR, through the register file address MUX, as described previously, and then requesting a read operation to load the contents of the register onto the bus.

For the arithmetic and logic instructions, the ALU (Arithmetic Logic Unit) performs the operation requested. It procures its operands from the two processor registers, DR (destination register) and SR (source register). These registers must be filled from the register file, using the destination field and the source field of the instruction as register numbers. The connections that allow the source and destination registers to be read have already been covered. Once read, the data can be transferred into the SR and DR registers, using their input connections to the bus. After the computation of the ALU is completed, the result is put on the bus, using the ALU output connection, and from there it can be loaded into the register file, as the destination register value.

The jump instruction requires data-path connections allowing the address field of the instruction to be loaded into the PC. As can be seen, the PC register has an input port connected to the bus, allowing it to receive the address, if it is placed on the bus. The address field can be placed on the bus, using the connection from the IR register to the bus, as discussed.

The conditional branch instructions perform the same type of transfer operation as the jump instruction. They, however, need circuitry allowing them to check the Z status flag. We see the Z register connected to the output of the ALU through a NOR gate. The NOR gate forms the NOR of all 16 bits of the ALU result. Remember that the NOR function produces a 1 if and only if all of its operands are 0. As a consequence, Z is loaded with 1, if the result is all zero, and otherwise it is loaded with a 0.

The input and output instructions use the BRIM machine I/O device. So, you can see that we have included data-path circuitry that allows us to get data into and out of an I/O device.

Before we leave the data-path in Figure 7.2, it is interesting to talk a little about the control bus. Remember, that the big "U-shaped" bus is the data bus. The control bus is composed of all of the control lines that operate the devices. We have two types of control lines: input control lines, that are used to send signals from the control unit to the device, and output control lines, or *status lines*, that send information from a device, back to the CU.

The input control lines for our BRIM machine are listed below.

1. Register file load (RLD)

2. Write to memory (MW)

3. I/O load (IOLD)

4. Load program counter (PCLD)

5. Load destination register (DRLD)

6. Load source register (SRLD)

7. Load instruction register (IRLD)

8. Load address register (ARLD)

9. Load Z flag (ZLD)

10. Register file enable (RE)

11. Enable memory (ME)

12. Enable I/O device (IOE)

13. Select program counter (SPC)

14. Select arithmetic logic unit (SALU)

15. Select instruction address (SAD)

16. Select register file address (SRF)

17. Select source operand (SS)

18. Select destination operand (SD)

19. Increment program counter (INPC)

20. ALU op-code; 3-bit (ALUOP)

The job of most of these lines should be fairly obvious, since we know what load and enable lines do. The "select" lines control tri-state switches, which, in turn, control data entry onto the bus. The exceptions are the SRF line, that controls the MUX that is used to supply an address to the register file, and the SS and SD lines that control the mode MUX, that places source or destination operands on the bus.

The BRIM machine has only a few status lines. These are listed below.

1. Instruction register contents; 16-bit (IRX)

2. Z flag contents (ZX)

The status lines send information to the CU, allowing it to make decisions. The information sent is the current instruction being worked on (the value stored in the IR register), and the results of the last arithmetic or logic operation (the value of the Z flag).

The control unit can only communicate with data path elements through the control bus. As such, it is restricted to the control and status signals sent, and received. To get information about the IR register and the Z register, their contents must be sent over the control bus as status signals. This is what is happening, with the status signals ZX and IRX: ZX is the value of the Z register, and IRX is the contents of the IR register. In summary, the distinction between the IR and the signal IRX is only minor; the IR is the register, and the IRX is the status line from the IR to the CU. A similar distinction exists between the Z register and the control bus line ZX.

If we examine Figure 7.2 carefully, and ask ourselves whether we know how to build each of the components of the data-path, the conclusion we should reach is that, yes, we can build, or have built all of the devices on the bus, except for the ALU and the mode MUX. So, we now need to talk about the design of these two components.

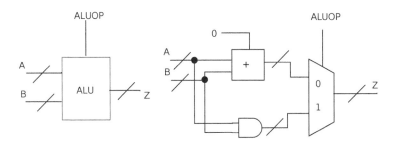

FIGURE 7.5 A simple add-AND ALU.

7.1.1.2 The ALU

An ALU is a device with operand inputs, result outputs, and a control input, which we call ALUOP. The interface to a simple ALU is shown in Figure 7.5. The leftmost diagram shows a device with operands A and B, result Z, and the control input ALUOP. This particular ALU is only capable of adding its operands, or producing the bit-wise AND of the two operands. The control line ALUOP determines which of the two operations is produced as the result.

The leftmost diagram of Figure 7.5 shows the implementation of the ALU. In its simplest form, an ALU is a multiplexer, connected to several *computational units*. Each computational unit computes one of the operations that the ALU is capable of performing. All computational units produce results, simultaneously. The ALUOP control lines control the MUX, selecting the result of one of the computational units as the final result, Z.

In our simple two-operation ALU, the ALUOP input is just one bit: if 0, we perform an addition operation, and if 1, we perform a bit-wise AND operation. So, in the implementation, the MUX has Option 0 connected to an adder that adds the A operand, the B operand, and a carry-in of 0. Option 1 is connected to an AND gate array that forms the bit-wise AND of operand A and operand B.

As the ALU becomes more complex, this basic design for an ALU needs a bit more organization. To demonstrate this, we now turn to the ALU of the BRIM machine. The BRIM ALU supports two arithmetic operations, and four logical operations: addition, subtraction, identity, AND, OR, and NOT. To help organize these operations, we split them into two separate categories: arithmetic and logic. We handle each of these two categories with two separate devices: an Arithmetic Unit (AU) and a Logic Unit (LU). Each of these two units can be thought of as a small ALU, in its own right, that specializes in operations of a specific type, and is structured in a similar way to our simple ALU of Figure 7.5.

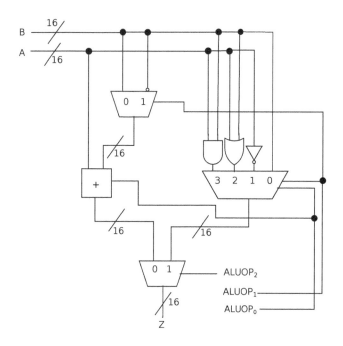

FIGURE 7.6 The BRIM ALU.

The BRIM ALU is shown in Figure 7.6. Starting from the bottom of the figure, we see an output MUX which chooses between the output of two devices: on the right is the LU, and on the left is the AU. The LU is composed of a MUX that chooses between the outputs of several computational units. From right to left we see an identity computational unit, an inverter array, an OR gate array, and an AND gate array. Each of the computational units are connected to the A and B inputs of the ALU.

The AU is centered around a multi-bit adder. Its A operand is connected to the A input of the ALU. Its carry-in is connected to a control line, $ALUOP_0$, and its B operand is connected to a multiplexer, that can supply either the B input of the ALU, or the B input inverted.

The control lines of the ALU form the ALU op-code, $ALUOP$. Alternately, we might say that the ALUOP is a 3-bit op-code composed of several fields.

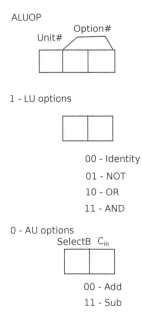

ALUOP

Unit# Option#

1 - LU options

00 - Identity
01 - NOT
10 - OR
11 - AND

0 - AU options

SelectB C$_{in}$

00 - Add
11 - Sub

FIGURE 7.7 ALUOP format.

Each field controls a different part of the ALU. Figure 7.7 shows how the ALUOP is split into fields. The top diagram shows that the ALUOP is split into two fields: a unit number and an option number. The unit number is used to choose between the AU and the LU. Once you have chosen the unit number, you are restricted to the operations that particular unit can perform. The option number then chooses between the operations available to the unit.

The middle diagram of Figure 7.7 shows the options for the LU. The two bits of the option number would choose one of the four options as output for the LU. The bottom diagram shows the options for the AU. The option number is now split into two separate bits: one bit gives the value for the carry-in, and one bit is used to control the select line on the MUX controlling the B input to the adder. A selection value of 0 uses the straight B value as input, and a selection value of 1 uses the inverse of B as the input.

Let us do a couple of examples. To perform an addition, we would specify that the AU should be used (unit number 0), that the carry-in should be 0, and that the B operand should be uninverted (selection 0). The resulting ALUOP would then be 000. The subtraction operation would add the two's compliment of the B operand to the A operand. This is done by specifying a unit number of 0, inverting the B operand, and using a carry-in value of one, giving 011 as the ALUOP value. (Subtraction is covered, later, in detail in Section 8.2.2.2.) An OR operation would use the LU (Unit 1), and its option number on the MUX is 2 (10, in binary), yielding an ALUOP value of 110.

7.1.1.3 The Mode MUX

Another unexplained component of Figure 7.2 is the mode multiplexer. Remember that the BRIM machine has three addressing modes: direct mode, register direct mode, and immediate mode. Direct mode is available only on the load, store, and branch instructions. With only a few instructions that use direct mode, it makes sense to handle the choice of this mode as part of the instruction control circuitry. However, with register direct mode and immediate mode, there are so many instructions that use these modes, that, to simplify the complexity of the machine, we handle the choice of these modes in a separate special circuit: the mode multiplexer.

The mode MUX, marked "mode" in Figure 7.2, places a source operand on the bus. It has, basically, two data inputs: a value coming from the register file and the current instruction. If the addressing mode of the source operand is register direct, it would put the value coming from the register file on the bus. If the addressing mode of the source operand was immediate mode, the source field from the instruction would be put on the bus.

To make the choice between the register value or the immediate value, the mode MUX must examine the source operand addressing mode bit, IR_7. If this bit is a 0, then register direct mode is indicated, and if it is a 1, then immediate mode is indicated.

The mode MUX has two control inputs, SS and SD. The SS input is activated when the source operand is being put on the bus. The internal circuitry then decides whether to put the register value or the immediate value on the bus. The SD line is activated when the destination operand is being read. Remember that the destination is always a register direct operand, and so when the SD is asserted, the register file value is always placed on the bus, without consulting the addressing mode bit.

Figure 7.8 shows the structure of the mode MUX. Three tri-state switches are used to control the flow of data onto the bus connection: the top two control for the source operand, and the bottom switch controls the destination. The input to the bottom switch is the value of the register file, and is controlled by the input SD. The top two switches are controlled by SS. The top switch is triggered if SS is asserted, and bit IR_7, the addressing mode bit for the source operand, is cleared. This switch puts the register value on the bus. The

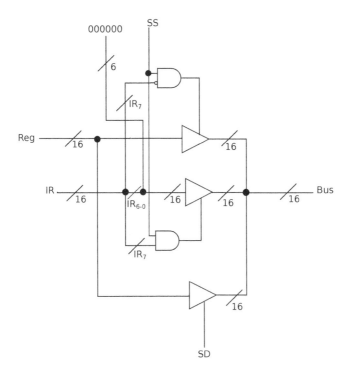

FIGURE 7.8 Mode MUX structure.

middle switch is triggered if SS is asserted, and the addressing mode bit is set, and places the immediate value on the bus.

There is still one point that needs clarification in Figure 7.8. The immediate value, coming from the instruction register, is only 7 bits. To form a data value out of it, it must be extended to 16 bits. You will notice that before the immediate value enters the switch, it is combined with six bits of zeros. These bits become the high-order bits of the 16-bit word, and the immediate value becomes the rightmost seven bits of a full 16-bit word.

Once we have the mode MUX incorporated into the data-path, we are relieved of worrying about whether an operand is register direct, or immediate value. If we wish to read the destination operand, we simply trigger the SD line, and the value of the destination register appears on the bus, assuming that the register file has been addressed correctly. If we wish to read the source operand, we simply trigger the SS line. The mode MUX then automatically figures out whether the immediate value or the register direct value should be placed on the bus.

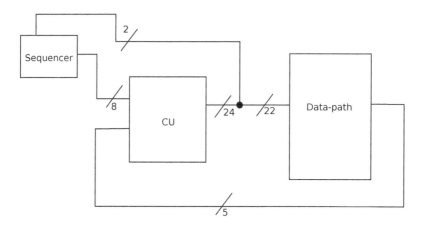

FIGURE 7.9 RIM processor structure.

7.1.2 The RIM Control Unit

We are designing our processor in the same way that we designed algorithmic circuitry in Section 5.3. Already, we have designed the data-path. We now design a control unit. The control unit will use a sequencer, just as in Section 5.3. Figure 7.9 is an abstract diagram of our processor, and shows the relationship between the CU and the other two components of the processor. The sequencer sends 8 timing signals to the CU. In addition to timing signals, the CU receives 5 status signals from the data-path: 4 bits for the IRX signal, and 1 bit for the ZX signal. Using these input signals, the CU calculates signals to move the machine into the next state. It sends out 24 signals: the 22 signals controlling the bus, and 2 bits that will control the sequencer.

To design the CU we must first describe, in detail, what it does. We know that it implements the machine cycle, but we need to be more specific. We also know that the machine cycle consists of three steps: fetch the next instruction, decode the instruction, and execute the instruction. Let us now examine how those steps might be implemented using our data-path.

7.1.2.1 Fetching an Instruction

To fetch an instruction, the next instruction in the program must be brought in from memory, and placed into the IR register. The processor knows that the address of the next instruction is stored in the PC. So, we could do an instruction fetch as follows

$$AR \leftarrow PC$$
$$IR \leftarrow M[AR], PC \leftarrow PC + 1 \tag{7.1}$$

In the first micro-instruction, we place the PC on the data bus, and trigger the AR register to load. Once we have loaded the PC into the address register, we can address memory, in the second micro-instruction, and move the result that is read, via the data bus, into the IR. Simultaneously we increment the PC, so that it will be pointing to the correct next instruction, for the next iteration of the machine cycle. The sequence is performed in two clock cycles.

7.1.2.2 Decoding an Instruction

When decoding an instruction, the processor examines the op-code in the IR, using the IRX status lines, to determine how to properly execute the instruction. In terms of actions on the data path, nothing is actually done in the decoding step. The result of the decoding is simply a change of sequencer state in the control unit.

7.1.2.3 Executing an Instruction

This stage of the machine cycle differs from instruction to instruction. For instance, for an addition instruction, two operands must be fetched from the register file, added, and the result must be written back to the register file. On the other hand, for a jump instruction, an address must be loaded into the PC. So, we need to examine the actions performed by each instruction, individually.

Executing the *movR/M* Instruction

The *movR/M* instruction is the most complex instruction in the RIM instruction set. This is mostly because of the complicated addressing modes associated with it; the source operand can come from the register file, the memory unit, or from the instruction itself. Regardless of the mode, the source operand must be loaded into the SR processor register. From there the operand can be moved through the ALU, onto the bus, and from there into the register file.

Let us examine the fetching of the source operand. If the source operand is given using direct mode, we would get the following micro-code.

$$AR \leftarrow IR_{7-0}$$
$$SR \leftarrow M[AR] \tag{7.2}$$

If the source is given in register direct mode, or immediate mode, we get the following micro-code.

$$SR \leftarrow \text{if } IR_7 \text{ then } IR_{6-0} \text{ else } R[IR_{6-4}] \tag{7.3}$$

This micro-instruction either loads the source field directly into the SR register, or uses the source register number to address the register file, R, and loads the value fetched into SR. An interesting, and significant, point is that this fetch takes only one clock cycle, whereas the fetch for direct mode takes two cycles. The problem is that to keep timing simple, we would like the two types of fetches to take the same amount of time. To solve this problem, we will stall the fetch for the register direct fetch for one clock cycle, making the two sequences the same length.

Once the operand has been loaded into the SR register, the micro-code is the same for all types of move instructions.

$$R[IR_{10-8}] \leftarrow SR \tag{7.4}$$

The full sequence will take three clock cycles.

Executing the *store* Instruction

A store instruction must fetch an operand from the register file. The operand is given by the destination field, in register direct mode. Once fetched, the operand can be loaded into memory. All this is done by the following micro-code sequence.

$$\begin{aligned} AR &\leftarrow IR_{7-0} \\ M[AR] &\leftarrow R[IR_{10-8}] \end{aligned} \tag{7.5}$$

We have a two-cycle sequence that first loads the address field from the instruction into the AR register. Then, in the second cycle, the address and the register number from the instruction are used to fetch the effective operand from the register file, and write it to memory.

Executing the Arithmetic and Logic Instructions

Arithmetic and logic instructions fetch their operands from registers. After the operands are fetched, they are placed in the SR and DR registers. The ALU is then asked to perform the appropriate operation, and the result is written back to the register file.

For all arithmetic and logic instructions, the operand fetch is the same.

$$\begin{aligned} SR &\leftarrow \text{if } IR_7 \text{ then } IR_{6-0} \text{ else } R[IR_{6-4}] \\ DR &\leftarrow R[IR_{10-8}] \end{aligned} \tag{7.6}$$

For loading the SR register, the addressing mode bit is consulted. Depending on its value, either the immediate value is loaded, or the source register is loaded.

After the operand fetch, another micro-instruction is used to write the result to the destination register, and set the Z flag to the NOR of all of the bits of the result. This last micro-instruction is shown for all arithmetic and logic operations, below.

- Add.

$$R[IR_{10-8}] \leftarrow DR + SR, Z \leftarrow \overline{\bigvee_{i=0}^{15}(DR + SR)_i} \tag{7.7}$$

- Sub.

$$R[IR_{10-8}] \leftarrow DR - SR, Z \leftarrow \overline{\bigvee_{i=0}^{15}(DR - SR)_i} \tag{7.8}$$

- AND.

$$R[IR_{10-8}] \leftarrow DR \wedge SR, Z \leftarrow \overline{\bigvee_{i=0}^{15}(DR \wedge SR)_i} \tag{7.9}$$

- OR.

$$R[IR_{10-8}] \leftarrow DR \vee SR, Z \leftarrow \overline{\bigvee_{i=0}^{15}(DR \vee SR)_i} \tag{7.10}$$

- NOT.

$$R[IR_{10-8} \leftarrow \overline{DR}, Z \leftarrow \overline{\bigvee_{i=0}^{15} \overline{DR}_i} \tag{7.11}$$

Executing the Branch Instructions

The branch instructions use an operand that is given in immediate mode. As a consequence there is no operand fetch. All branch instructions load the address field of the instruction into the PC. That is what the following micro instruction does.

$$PC \leftarrow IR_{7-0} \tag{7.12}$$

The difference between the *bz*, *bnz*, and *jump* instructions is just a question of when this branch micro-instruction is performed. For the *bz* and *bnz* instructions, the branch micro-instruction is triggered by the value of the Z flag, and for the *jump* instruction, the branch micro-instruction is always triggered.

Executing the I/O Instructions

The *in* instruction moves the current value of the I/O device into the destination register. The *out* instruction moves the value of the source register, or an immediate value, into the I/O device's device register.

To implement these two instructions with our data path, we simply write micro-instructions that perform the data transfers via the data bus. So, for the input instruction, we get the following micro-instruction.

$$R[IR_{10-8}] \leftarrow D_{in} \qquad (7.13)$$

For the output instruction we get the following.

$$D_{out} \leftarrow \text{if } IR_7 \text{ then } IR_{6-0} \text{ else } R[IR_{6-4}] \qquad (7.14)$$

7.1.2.4 The CU Behavioral Description

Let us now take the micro-code we have written for the machine cycle, and convert it into a full behavioral description of the CU. Before we can do this, however, we must spend a little time on the detailed structure of the CU.

Figure 7.9 shows how the CU is wired up to the data-path and the sequencer. It is indicated that the CU receives 5 bits of input from the data-path, and 8 bits of input from the sequencer. The CU outputs 24 bits; 22 bits are sent to the data-path, and 2 bits are used to control the sequencer. We have not yet identified these input and output bits for the CU. And, before we can proceed, we must remedy this.

Figure 7.10 shows a magnification of the control unit. The core of the control unit is a combinational circuit, marked as "control," on the diagram. The control circuit outputs the 22 control lines to the data path, and two control lines to the sequencer register. As can be seen in the top left of Figure 7.10, the lines sent to the sequencer register are the usual increment, and clear lines, for the standard register from Section 3.2.5.4.

The control circuit receives input from the control bus, and from the sequencer. From the sequencer it receives the usual timing lines. For the RIM machine, the counter for the sequencer is a 3-bit register, and so the time decoder will produce eight time step signals, T_0 through T_7. From the control bus, the control circuit receives the ZX signal, the contents of the Z flag, and IR_{15-12}, the 4-bit op-code from the IR register. To simplify wiring, however, the op-code is not fed directly into the control circuit. Instead, it is sent through a decoder, the op-code decoder. This decoder produces trigger lines corresponding to the op-code. These trigger lines are named OP_k, where k is

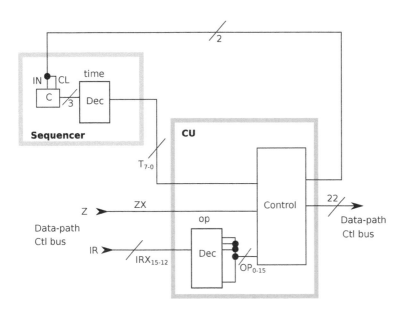

FIGURE 7.10 The CU, sequencer, and op-code decoder.

an op-code. For example, if OP_5 is turned on, this would indicate that the current instruction has op-code 5, corresponding to the *or* instruction.

You may notice that the op-code decoder produces some signals that are never used. For instance op_{15} indicates that the op-code is fifteen. But, we do not have a machine instruction with op-code fifteen, and so this trigger line would not be used by the control circuit. Similarly, the time decoder produces time signals that are not used. We shall see shortly that our machine only uses time signals T_0 through T_4.

To summarize the information in Figure 7.10, the control circuit uses the time step, from the time decoder, and the op-code signals, from the op-code decoder, to determine what to do. The value of the Z register is a less important input, and determines only what happens with conditional branches.

We can now build a flowchart for our CU. This flow chart has been drawn in Figure 7.11. It shows the complete machine cycle for the BRIM machine. The cycle is composed of *phases*. Each phase has been given a name. For instance, the top phase is called F_0, indicating that it is the first phase of the fetch step of the machine cycle. Some of the phases are decision diamonds, and some of

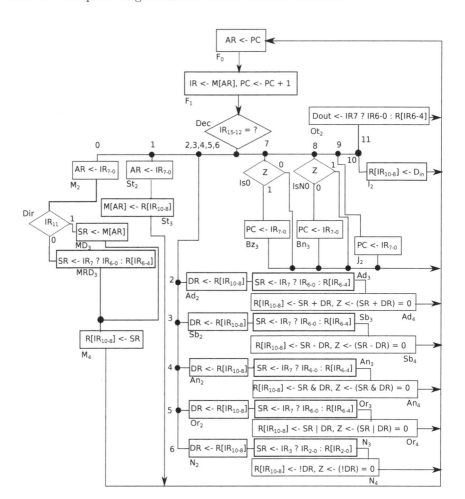

FIGURE 7.11 RIM CU flowchart.

them are action rectangles. In Chapter 5, when constructing an algorithmic circuit, we made each phase into a step performed at a separate clock cycle. For our purposes now, we perform decisions using control circuitry, and so only action phases will be performed on the clock cycles.

You might wonder how we decided on the subscripting for the phases. The answer is that we chose subscripts reflecting the timing of the operations. For instance, the phase F_0 has a subscript 0, indicating that this micro-operation is performed on the very first clock cycle of the machine cycle. Similarly, MD_2, with a subscript of 2, would be performed on the third clock cycle.

To explain Figure 7.11 we begin at the top, with the fetch step of the machine cycle. The instruction fetch is done in two phases, F_0 and F_1. In the first phase, the AR register is loaded with the address currently in the PC. In the second phase, the instruction at that address is fetched into the IR register, and the PC is updated.

The decode step of the machine cycle is implemented with the phase Dec. This is a decision phase, in which we use the instruction op-code to send control to one of eleven branches of the flowchart. Each branch is a sub-chart, which implements the execution of the particular machine instruction.

Chart Branch 0 corresponds to the chart branch handling the $moveR/M$ instruction (op-code 0). In phase M_2, the instruction address field is loaded into the AR register. The decision phase Dir checks the destination addressing mode bit, to determine if the source operand is in direct mode. If the source is in direct mode, MD_3 loads the SR register from memory. If the source is in register direct mode, or immediate mode, phase MRD_3 loads the SR register from the register file, or from the instruction register, respectively. Once the SR register is loaded, phase M_4 transfers the contents of the SR register to the register file.

Chart Branch 1 implements the store instruction. It consists of phase St_2, that loads the instruction address field into the AR register, and phase St_3, that transfers the value from the register file into memory.

Moving on to the chart branch for op-codes 2, 3, 4, 5, and 6, these branches implement the ALU instructions. Each instruction is similar to the others, and so we examine only the add instruction. On the chart branch labeled 2, we first load the two operands from the register file into the DR and SR registers, in phases Ad_2 and Ad_3. Then, in phase Ad_4, we perform the add operation, writing the result back to the register file, and the status to the Z register.

For the two conditional branches, op-codes 7 and 8, the process is identical, except for the condition evaluated. For the bz instruction, in phase Bz_2 we check to see if the Z flag is set, and if so, in phase Bz_3 we load the instruction address field into the PC. If the Z flag is cleared, we do nothing. For the bnz instruction we load the PC if the Z flag is cleared, rather than set.

The sub-chart for the jump instruction is simple. In phase J_2, the address field is loaded into the PC, completing the execution of the machine instruction.

The chart branches that handle the I/O instructions, op-codes 10 and 11, simply do the transfer to or from the device registers. For the input instruction, the transfer is done in phase I_2, and for the output instruction, the transfer is done in phase Ot_2. Notice that phase Ot_2 transfers either an immediate value, coming from the instruction register, or a register direct value, coming out of the register file.

After the execution of each machine instruction, our flowchart returns to phase F_0, to start a new iteration of the machine cycle. In this iteration, however, the machine instruction executed will be from the next word in memory, or from the target of a branch, if a branch was executed.

We have now defined our CU in terms of micro-operations that are performed in different phases. The problem is that these phases are high-level specifications. In order to build the CU we need to define what a phase is, in terms of the low-level inputs to the CU. As seen in Figure 7.10, the low-level inputs to the control unit are the timing signals, T_0 through T_7, the op-code signals coming from the op-code decoder, OP_0 through OP_{15}, and the signal ZX, coming from the Z flag.

So, let us now try to figure out how the control circuit will know what phase it is in, given the low-level status input. We start with phase F_0. Clearly, F_0 shall be performed at time step T_0. As a matter of fact, this is sufficient information, and so F_0 is equivalent to T_0.

As another example, consider the phase MRD_3. First we compute the timing signal for MRD_3. We have already mentioned that we will perform each action phase on a separate time signal. So, F_0 will be performed at time step T_0, F_1 will be performed at time step T_1, M_2 will be executed at time step T_2, etc. As a matter of fact, we have already mentioned that the index on each phase name indicates its time step. And so, the time step for MRD_3 will be T_3.

There is more terminal phases to phase MRD_3, however. If we ask ourselves, how do we arrive at phase MRD_3, we come up with several necessary conditions. First, we must make it through phase D_1, and take the leftmost branch in the flowchart. This implies that the op-code must be 0, or equivalently, OP_0 is set. Then we must make it through Dir, which guaranties that $\overline{IRX_{11}}$. In total, the phase $MRD_3 \equiv T_3 \cdot OP_0 \cdot \overline{IRX_{11}}$.

We can continue this process, and assign input conditions to all of the phases, using the index of the phase to determine the time step, and following through all decision phases to determine conditions on IRX and ZX. Table

TABLE 7.1   Phases as control signals.

Phase	Control Signals	Sequencer Control
F_0	T_0	$C \leftarrow C + 1$
F_1	T_1	$C \leftarrow C + 1$
M_2	$T_2 \cdot OP_0$	$C \leftarrow C + 1$
St_2	$T_2 \cdot OP_1$	$C \leftarrow C + 1$
Ad_2	$T_2 \cdot OP_2$	$C \leftarrow C + 1$
Sb_2	$T_2 \cdot OP_3$	$C \leftarrow C + 1$
An_2	$T_2 \cdot OP_4$	$C \leftarrow C + 1$
Or_2	$T_2 \cdot OP_5$	$C \leftarrow C + 1$
N_2	$T_2 \cdot OP_6$	$C \leftarrow C + 1$
Bz_2	$T_2 \cdot OP_7 \cdot ZX$	
Bn_2	$T_2 \cdot OP_8 \cdot \overline{ZX}$	
$Bz_2 \vee \overline{Bz_2}$	$T_2 \cdot OP_7$	$C \leftarrow 0$
$Bn_2 \vee \overline{Bn_2}$	$T_2 \cdot OP_8$	$C \leftarrow 0$
J_2	$T_2 \cdot OP_9$	$C \leftarrow 0$
I_2	$T_2 \cdot OP_{10}$	$C \leftarrow 0$
Ot_2	$T_2 \cdot OP_{11}$	$C \leftarrow 0$
MD_3	$T_3 \cdot OP_0 \cdot IRX_{11}$	$C \leftarrow C + 1$
MRD_3	$T_3 \cdot OP_0 \cdot \overline{IRX_{11}}$	$C \leftarrow C + 1$
St_3	$T_3 \cdot OP_1$	$C \leftarrow 0$
Ad_3	$T_3 \cdot OP_2$	$C \leftarrow C + 1$
Sb_3	$T_3 \cdot OP_3$	$C \leftarrow C + 1$
An_3	$T_3 \cdot OP_4$	$C \leftarrow C + 1$
Or_3	$T_3 \cdot OP_5$	$C \leftarrow C + 1$
N_3	$T_3 \cdot OP_6$	$C \leftarrow C + 1$
M_4	$T_4 \cdot OP_0$	$C \leftarrow 0$
Ad_4	$T_4 \cdot OP_2$	$C \leftarrow 0$
Sb_4	$T_4 \cdot OP_2$	$C \leftarrow 0$
An_4	$T_4 \cdot OP_4$	$C \leftarrow 0$
Or_4	$T_4 \cdot OP_5$	$C \leftarrow 0$
N_4	$T_4 \cdot OP_6$	$C \leftarrow 0$

7.1 summarizes the results we would produce in the first two columns. Notice that Table 7.1 contains no entries for the decision phases. These phases have now been incorporated into the control signals of the phases that follow them in the flowchart.

Another topic of importance is the control of the sequencer, and its counter C. As can be seen from Figure 7.11, after each phase, the next phase is at the next time step, until you arrive at the bottom of the flowchart. As an example, once phase Ad_2 is completed, which is done at time step T_2, the next phase to be performed is phase Ad_3, at time step T_3. To move from time step T_2 to time step T_3, you simply increment the counter in the sequencer, C, from 2 to 3. This is true of every phase in Figure 7.11, except the phases that terminate branches. For those phases, the next phase to be performed is phase F_0, at time step T_0. In other words, for these terminal phases, the counter C should be cleared. Returning to Table 7.1, and examining the leftmost column, this is exactly what is being described. The last column gives the control for the sequencer. You can verify that for most phases, the sequencer is incremented. For phases that terminate a flowchart branch, the counter is cleared.

Notice that in Table 7.1, strangely, we have not included sequencer control for the phases that handle cases in which conditional branches are taken: Bz_2, and Bn_2. This is because we are combining sequencer control with the cases in which conditional branches are not taken. We have added two very strange, new phases, that are not in Figure 7.11: $Bz_2 \vee \overline{Bz_2}$ and $Bn_2 \vee \overline{Bn_2}$. They simply state that, at time step T_2, for op-codes OP_7 or OP_8, always reset the sequencer, regardless of whether or not the PC is loaded with a new address.

We are now ready for a complete behavioral description of the CU. The micro-program in Tables 7.2 and 7.3 is the behavioral description. The first two columns give the phase and the micro-instruction that is performed at that phase. Little is new here, although you will notice that we have included the micro-operations that control the sequencer, in the micro-instructions.

The third column of Tables 7.2 and 7.3 is new. What we have done in this column is to indicate all control signals that need to be sent to the data-path, or sequencer, to perform the micro-instruction indicted for each phase. We developed this column, much in the same way as we wired control for RTL circuits in Chapter 5. The data-path diagram in Figure 7.2 was essential in this process, and we will demonstrate the technique with a few examples.

We begin with the phase F_0. The first micro-operation of this phase loads the PC into the AR register. Examining the data-path, we see that to do this we first put the PC on the bus. This is accomplished by opening the tri-state

TABLE 7.2 Micro-program for the BRIM machine.

Phase	Micro-instruction	Control Output
F_0	$T_0 : AR \leftarrow PC, C \leftarrow C + 1$	ARLD, SPC, CIN
F_1	$T_1 : IR \leftarrow M[AR], PC \leftarrow PC + 1,$ $C \leftarrow C + 1$	IRLD, ME, PCIN, CIN
M_2	$T_2 \cdot OP_0 : AR \leftarrow IR_{7-0}, C \leftarrow C + 1$	ARLD, SAD, CIN
MD_3	$T_3 \cdot OP_0 \cdot IRX_{11} : SR \leftarrow M[AR],$ $C \leftarrow C + 1$	SRLD, ME, CIN
MRD_3	$T_3 \cdot OP_0 \cdot \overline{IRX_{11}} :$ $SR \leftarrow$ if IR_7 then IR_{6-0} else $R[IR_{6-4}], C \leftarrow C + 1$	SRLD, RE, SS, CIN
M_4	$T_4 \cdot OP_0 : R[IR_{10-8}] \leftarrow SR, C \leftarrow 0$	RLD, SRF, SALU, CCL, ALUOP = 100
St_2	$T_2 \cdot OP_1 : AR \leftarrow IR_{7-0}, C \leftarrow C + 1$	ARLD, SAD, CIN
St_3	$T_3 \cdot OP_1 : M[AR] \leftarrow R[IR_{10-8}],$ $C \leftarrow 0$	MW, RE, SRF, SD, CCL
Ad_2	$T_2 \cdot OP_2 : DR \leftarrow R[IR_{10-8}],$ $C \leftarrow C + 1$	DRLD, RE, SD, SRF, CIN
Ad_3	$T_3 \cdot OP_2 :$ $SR \leftarrow$ if IR_7 then IR_{6-0} else $R[IR_{6-4}], C \leftarrow C + 1$	SRLD, RE, SS, CIN
Ad_4	$T_4 \cdot OP_2 : R[IR_{10-8}] \leftarrow SR + DR,$ $Z \leftarrow (SR + DR) = 0, C \leftarrow 0$	RLD, SRF, SALU, ZLD, ALUOP = 000, CCL
Sb_2	$T_2 \cdot OP_3 : DR \leftarrow R[IR_{10-8}],$ $C \leftarrow C + 1$	DRLD, RE, SD, SRF, CIN
Sb_3	$T_3 \cdot OP_3 :$ $SR \leftarrow$ if IR_7 then IR_{6-0} else $R[IR_{6-4}], C \leftarrow C + 1$	SRLD, RE, SS, CIN
Sb_4	$T_4 \cdot OP_3 : R[IR_{10-8}] \leftarrow SR - DR,$ $Z \leftarrow (SR - DR) = 0, C \leftarrow 0$	RLD, SRF, SALU, ZLD, ALUOP = 011, CCL,
An_2	$T_2 \cdot OP_4 : DR \leftarrow R[IR_{10-8}],$ $C \leftarrow C + 1$	DRLD, RE, SD, SRF, CIN
An_3	$T_3 \cdot OP_4 :$ $SR \leftarrow$ if IR_7 then IR_{6-0} else $R[IR_{6-4}], C \leftarrow C + 1$	SRLD, RE, SS, CIN
An_4	$T_4 \cdot OP_4 : R[IR_{10-8}] \leftarrow SR \wedge DR,$ $Z \leftarrow (SR \wedge DR) = 0, C \leftarrow 0$	RLD, SRF, SALU, ZLD, ALUOP = 111, CCL
Or_2	$T_2 \cdot OP_5 : DR \leftarrow R[IR_{10-8}],$ $C \leftarrow C + 1$	DRLD, RE, SD, SRF, CIN
Or_3	$T_3 \cdot OP_5 :$ $SR \leftarrow$ if IR_7 then IR_{6-0} else $R[IR_{6-4}], C \leftarrow C + 1$	SRLD, RE, SS, CIN
Or_4	$T_4 \cdot OP_5 : R[IR_{10-8}] \leftarrow SR \vee DR,$ $Z \leftarrow (SR \vee DR) = 0, C \leftarrow 0$	RLD, SRF, SALU, ZLD, ALUOP = 110

TABLE 7.3 Micro-program for the BRIM machine (Cont.).

Phase	Micro-instruction	Control Output
N_2	$T_2 \cdot OP_6 : DR \leftarrow R[IR_{10-8}],$ $\quad C \leftarrow C + 1$	DRLD, RE, SD, SRF, CIN
N_3	$T_3 \cdot OP_6 :$ $\quad SR \leftarrow$ if IR_7 then IR_{6-0} $\quad$ else $R[IR_{6-4}], C \leftarrow C + 1$	SRLD, RE, SS, CIN
N_4	$T_4 \cdot OP_6 : R[IR_{10-8}] \leftarrow \overline{DR},$ $\quad Z \leftarrow \overline{DR} = 0, C \leftarrow 0$	RLD, SRF, SALU, ZLD, ALUOP $= 101$, CCL
Bz_2	$T_2 \cdot OP_7 \cdot ZX : PC \leftarrow IR_{7-0}$	PCLD, SAD
$Bz_2 \vee \overline{Bz_2}$	$T_2 \cdot OP_7 : C \leftarrow 0$	CCL
Bn_2	$T_2 \cdot OP_8 \cdot \overline{ZX} : PC \leftarrow IR_{7-0}$	PCLD, SAD
$Bn_2 \vee \overline{Bn_2}$	$T_2 \cdot OP_8 : C \leftarrow 0$	CCL
J_2	$T_2 \cdot OP_9 : PC \leftarrow IR_{7-0}, C \leftarrow 0$	PCLD, SAD, CCL
I_2	$T_2 \cdot OP_{10} : R[IR_{10-8}] \leftarrow D_{in} : C \leftarrow 0$	RLD, SRF, IOE, CCL
Ot_2	$T_2 \cdot OP_{11} :$ $\quad D_{out} \leftarrow$ if IR_7 then IR_{6-0} $\quad$ else $R[IR_{6-4}], C \leftarrow 0$	SS, RE, IOLD, CCL

switch connecting the PC to the bus, which is, in turn, accomplished by setting the SPC line to a 1. Then we must open the AR register to receive the value on the bus. This is done by triggering the ARLD line on the AR register. The second micro-operation performed in F_0 increments the sequencer counter, C. This would be done by triggering the increment pin on the C register, which we have called CIN. These three lines, ARLD, SPC, and CIN, are the lines listed in the third column for the phase F_0.

In the third column of Tables 7.2 and 7.3 we list all control pins that are triggered to perform the micro-instruction. You might wonder what happens to all of the other pins that are not listed. The answer is that if a pin is not listed, it is held at 0, for that clock cycle. As a consequence, only the devices connected to the control lines listed will be activated. All other devices will be disabled.

The next phase we will examine is Bz_2. The first micro-operation performed in the phase is one that transfers the address field of the IR register into the PC. Examining the data-path in Figure 7.2, we observe that the IR is connected to the bus via a connection controlled by the SAD line. So, to perform this micro-operation, we would trigger the SAD line and the PCLD line to open the PC up to receive the address. Finally, we clear the C counter by triggering its clear line, called CCL.

As a final example, consider the phase Ad_4. The first micro-operation of this phase adds the contents of the SR and DR registers, and then transfers the result into the register file, to the destination register. To start with, the result of the addition, produced by the ALU, must be put on the bus. This is done by triggering the SALU line, that controls a tri-state switch. Next the register file must be properly addressed. Examining Figure 7.2, we see that there are two addresses used for the register file, controlled by a MUX. The address can either be Option 0, the source register, or Option 1, the destination register. In this case we want the destination register to be used, and so we would set the line SRF, controlling the MUX, to a 1. With correct addressing, we would then open the register file to receive the arithmetic result, by triggering the load line, RLD.

There is one point that we have not covered yet, concerning the phase Ad_4. And, that is that we need to tell the ALU what operation to perform. In this case, we need to tell it to do an addition operation. Instructing the ALU to perform a particular operation is the job of the ALUOP control input, shown in Figure 7.2. The ALUOP is a 3-bit control input, called the ALU op-code, which is described in Section 7.1.1.2. It instructs the ALU to perform one of a fixed set of arithmetic or logic operations. The op-codes are as follows: 011 for subtraction, 111 for AND, 110 for OR, and 101 for NOT. There is also an operation, which we will call the *identity* operation, used in phase M_4, which simply transfers its operand through the ALU, unaltered. It has an ALU op-code of 100.

7.1.2.5 The Control Circuitry

To complete the design of the CU, we must transform our behavioral RTL description into circuitry. We have done jobs like this before, in Chapter 5. And, we have already started the work required in Table 7.2. As it turns out, the third column of this table contains most of the information required for building the circuitry.

As you are already aware, to construct a combinational circuit, we usually require Boolean equations for the outputs, in terms of the inputs. In the case of the CU, these outputs are the control lines of the control bus, and the inputs are the IRX, ZX, OP, and T lines, as shown in Figure 7.10. We often think of Table 7.2 as inverted equations. The table tells us what the outputs are, for a given set of inputs. To build Boolean equations, we need to know what the inputs are for a given output. In other words, we need to invert the table.

The technique for inverting the table is fairly straightforward. We demonstrate this by example. Let us determine the equation of the CU output SRLD.

TABLE 7.4   Control circuit equations for the BRIM machine.

Output Signal	Formula
RLD	$T_2 \cdot OP_{10} + T_4(OP_0 + OP_2 + OP_3 + OP_4 + OP_5 + OP_6)$
MW	$T_3 \cdot OP_1$
$IOLD$	$T_2 \cdot OP_{11}$
$PCLD$	$T_2(OP_7 \cdot ZX + OP_8 \cdot \overline{ZX} + OP_9)$
$DRLD$	$T_2(OP_2 + OP_3 + OP_4 + OP_5 + OP_6)$
$SRLD$	$T_3(OP_0 + OP_2 + OP_3 + OP_4 + OP_5 + OP_6)$
$IRLD$	T_1
$ARLD$	$T_0 + T_2(OP_0 + OP_1)$
ZLD	$T_4(OP_2 + OP_3 + OP_4 + OP_5 + OP_6)$
RE	$T_2(OP_2 + OP_3 + OP_4 + OP_5 + OP_6 + OP_{11})$
	$\quad + T_3(OP_0 \cdot \overline{IRX_{11}} + OP_1 + OP_2 + OP_3$
	$\quad + OP_4 + OP_5 + OP_6)$
ME	$T_1 + T_3 \cdot OP_0 \cdot IRX_{11}$
IOE	$T_2 \cdot OP_{10}$
SPC	T_0
$SALU$	$T_4(OP_0 + OP_2 + OP_3 + OP_4 + OP_5 + OP_6)$
SAD	$T_2(OP_0 + OP_1 + OP_7 \cdot ZX + OP_8 \cdot \overline{ZX} + OP_9)$
SRF	$T_2(OP_0 + OP_2 + OP_3 + OP_4 + OP_5 + OP_6 + OP_{10})$
	$\quad + T_4(OP_0 + OP_2 + OP_3 + OP_4 + OP_5 + OP_6)$
SS	$T_3 \cdot OP_{11}$
	$\quad + T_3(OP_0 \cdot \overline{IRX_{11}} + OP_2 + OP_3 + OP_4 + OP_5 + OP_6)$
SD	$T_2(OP_2 + OP_3 + OP_4 + OP_5 + OP_6) + T_3 \cdot OP_1$
$PCIN$	T_1
$ALUOP_0$	$T_4(OP_3 + OP_4 + OP_6)$
$ALUOP_1$	$T_4(OP_3 + OP_4 + OP_5)$
$ALUOP_2$	$T_4(OP_0 + OP_4 + OP_5 + OP_6)$
CIN	$T_0 + T_1$
	$\quad + T_2(OP_0 + OP_1 + OP_2 + OP_3 + OP_4 + OP_5 + OP_6)$
	$\quad + T_3(OP_0 + OP_2 + OP_3 + OP_4 + OP_5 + OP_6)$
CCL	$T_2(OP_7 + OP_8 + OP_9 + OP_{10} + OP_{11}) + T_3 \cdot OP_1$
	$\quad + T_4(OP_0 + OP_2 + OP_3 + OP_4 + OP_5 + OP_6)$

To determine this output, what we do is to examine the left column of Table 7.2. We write down every line on which we find that SRLD is asserted. So, we find that SRLD is asserted in phases MD_3, MRD_3, Ad_3, Sb_3, An_3, Or_3, and N_3. We then construct the equation by taking the OR of the inputs that indicate these phases. So, for example, MD_3 is indicated by $T_3 \cdot OP_0 \cdot IRX_{11}$. This would be part of the equation. Sb_3 is indicated by $T_3 \cdot OP_3$, and this would also be part of the equation. The full equation would be the following.

$$SRLD = T_3 \cdot OP_0 \cdot IRX_{11} + T_3 \cdot OP_0 \cdot \overline{IRX_{11}} + T_3 \cdot OP_2 + T_3 \cdot OP_3$$
$$+ T_3 \cdot OP_4 + T_3 \cdot OP_5 + T_3 \cdot OP_6$$

This simplifies to the following.

$$SRLD = T_3(OP_0 + OP_2 + OP_3 + OP_4 + OP_5 + OP_6)$$

We can continue this process, and develop equations for all of the CU outputs. The result would be the entries of Table 7.4.

If we take stock of where we currently are, we discover that we have, essentially, built a processor. We have fully described the data-path, shown in Figure 7.2. We have described the high-level structure of the processor, shown in Figure 7.9. And, now, we have described the control circuitry, by giving the output equations of Table 7.4. Table 7.4 is all we need to build the control circuit. We will not bore you by producing a circuit diagram; you have already produced circuit diagrams from such equations numerous times, substituting gates for the Boolean functions of the formula.

7.2   CONTROL FOR OTHER ARCHITECTURES

Our focus is on the register implicit architecture. But we have designed the RIM machine so that it can be relatively easily adapted to the other architectures we discussed in Chapter 6: the register machine, the accumulator machine, and the stack machine. We now spend some time discussing the processor designs for these other architectures. We will not delve into them in the same depth, but we will present enough information to develop a behavioral description of them.

7.2.1   Control for the Register Machine

The register machine is easily implemented on the RIM machine. This is because the RIM machine has been developed for the register implicit architecture, which is very similar, semantically, to the register architecture, and differs only in how operands are specified.

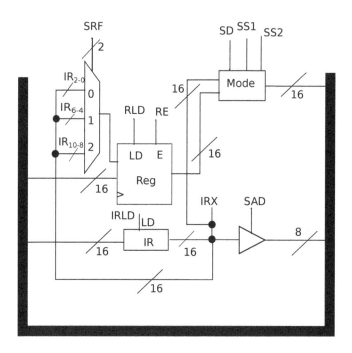

FIGURE 7.12 Register machine data-path modifications.

The data-path for the register machine is almost identical to that for the register implicit machine. The differences are associated with the use of a third operand field. This third field, the second source operand, as shown in Figure 6.11, comprises the bits IR_{3-0}. The second source operand can be either a register direct value, or an immediate mode value. The modifications to the data-path must, therefore, allow it to be used to address the register file, or allow it to be placed directly onto the bus. Modifications to the data-path are shown in Figure 7.12. We show only the parts of the bus that are modified; all devices remain unaltered, except for the register file and the IR.

The changes we see in Figure 7.12 include a new path for the second source operand, controlled by the mode MUX. The mode MUX now has three trigger lines: SD, which places the contents of the destination register on the bus, SS1, which places either the contents of the first source register, or the specified immediate value onto the bus, and SS2 that places the contents of the second source register, or the specified immediate value, on the bus. Also the address port on the register file is now connected to a larger MUX, with three choices: the first source operand, the second source operand, or the destination operand. The selection input controlling the MUX, SRF, is now 2 bits wide.

TABLE 7.5 Micro-code for the register machine.

Phase	Micro-instruction	Control Output
F_0	$T_0 : AR \leftarrow PC, C \leftarrow C + 1$	ARLD, SPC, CIN
F_1	$T_1 : IR \leftarrow M[AR], PC \leftarrow PC + 1,$ $C \leftarrow C + 1$	IRLD, ME, PCIN, CIN
L_2	$T_2 \cdot OP_0 : AR \leftarrow IR_{7-0},$ $C \leftarrow C + 1$	ARLD, SAD, CIN
L_3	$T_3 \cdot OP_0 : SR \leftarrow M[AR],$ $C \leftarrow C + 1$	SRLD, ME, CIN
L_4	$T_4 \cdot OP_0 : R[IR_{10-8}] \leftarrow SR,$ $C \leftarrow 0$	RLD, SALU, SRF = 10, CCL, ALUOP = 100
St_2	$T_2 \cdot OP_1 : AR \leftarrow IR_{7-0},$ $C \leftarrow C + 1$	ARLD, SAD, CIN
St_3	$T_3 \cdot OP_1 : M[AR] \leftarrow R[IR_{10-8}],$ $C \leftarrow 0$	MW, RE, SRF = 10, SD, CCL
Ad_2	$T_2 \cdot OP_2 : DR \leftarrow R[IR_{6-4}],$ $C \leftarrow C + 1$	DRLD, RE, SRF = 01, SS1, CIN
Ad_3	$T_3 \cdot OP_2 : SR \leftarrow R[IR_{2-0}],$ $C \leftarrow C + 1$	SRLD, RE, SRF = 00, SS2, CIN
Ad_4	$T_4 \cdot OP_2 : R[IR_{10-8}] \leftarrow SR + DR,$ $Z \leftarrow (SR + DR) = 0, C \leftarrow 0$	RLD, SRF = 10, SALU, ZLD, CCL, ALUOP = 000
Sb_2	$T_2 \cdot OP_3 : DR \leftarrow R[IR_{6-4}],$ $C \leftarrow C + 1$	DRLD, RE, SRF = 01, SS1, CIN
Sb_3	$T_3 \cdot OP_3 : SR \leftarrow R[IR_{2-0}],$ $C \leftarrow C + 1$	SRLD, RE, SRF = 00, SS2, CIN
Sb_4	$T_4 \cdot OP_3 : R[IR_{10-8}] \leftarrow SR - DR,$ $Z \leftarrow (SR - DR) = 0, C \leftarrow 0$	RLD, SRF = 10, SALU, ZLD, CCL, ALUOP = 011
An_2	$T_2 \cdot OP_4 : DR \leftarrow R[IR_{6-4}],$ $C \leftarrow C + 1$	DRLD, RE, SRF = 01, SS1, CIN
An_3	$T_3 \cdot OP_4 : SR \leftarrow R[IR_{2-0}],$ $C \leftarrow C + 1$	SRLD, RE, SRF = 00, SS2, CIN
An_4	$T_4 \cdot OP_4 : R[IR_{10-8}] \leftarrow SR \wedge DR,$ $Z \leftarrow (SR \wedge DR) = 0, C \leftarrow 0$	RLD, SRF = 10, SALU, ZLD, CCL, ALUOP = 111
Or_2	$T_2 \cdot OP_5 : DR \leftarrow R[IR_{6-4}],$ $C \leftarrow C + 1$	DRLD, RE, SRF = 01, SS1, CIN
Or_3	$T_3 \cdot OP_5 : SR \leftarrow R[IR_{2-0}],$ $C \leftarrow C + 1$	SRLD, RE, SRF = 00, SS2, CIN
Or_4	$T_4 \cdot OP_5 : R[IR_{10-8}] \leftarrow SR \vee DR,$ $Z \leftarrow (SR \vee DR) = 0, C \leftarrow 0$	RLD, SRF = 10, SALU, ZLD, CCL, ALUOP = 110

TABLE 7.6   Micro-code for the register machine (Cont.).

Phase	Micro-instruction	Control Output
N_2	$T_2 \cdot OP_6 : DR \leftarrow R[IR_{6-4}]$, $C \leftarrow C + 1$	DRLD, RE, SRF = 01, SS1, CIN
N_3	$T_3 \cdot OP_6 : SR \leftarrow R[IR_{2-0}]$, $C \leftarrow C + 1$	SRLD, RE, SRF = 00, SS2, CIN
N_4	$T_4 \cdot OP_6 : R[IR_{10-8}] \leftarrow \overline{DR}$, $Z \leftarrow \overline{DR} = 0, C \leftarrow 0$	RLD, SRF = 10, SALU, ZLD, CCL, ALUOP = 101
Bz_2	$T_2 \cdot OP_8 \cdot ZX : PC \leftarrow IR_{7-0}$	PCLD, SAD
$Bz_2 \vee \overline{Bz_2}$	$T_2 \cdot OP_8 : C \leftarrow 0$	CCL
Bn_2	$T_2 \cdot OP_9 \cdot \overline{ZX} : PC \leftarrow IR_{7-0}$	PCLD, SAD
$Bn_2 \vee \overline{Bn_2}$	$T_2 \cdot OP_9 : C \leftarrow 0$	CCL
J_2	$T_2 \cdot OP_7 : PC \leftarrow IR_{7-0}, C \leftarrow 0$	PCLD, SAD, CCL

Tables 7.5 and 7.6 give the specification for control of the register machine. Table 7.5 starts with the fetch phases, which are identical to those of the register implicit machine. The next section gives the micro-instructions for the load instruction. The load instruction replaces the *moveR/M* instruction of the register implicit machine. Unlike the move instruction, the load instruction implements only one addressing mode: direct addressing. Because of this, the micro-code is much simpler; in L_2 we load the address field into the AR register, in L_3 we fetch the data into the SR register, and in L_4 we move the data from the SR register into the destination register.

The micro-code for the store instruction on the register machine is no different than the same instruction on the register implicit machine. Moving on to the ALU instructions, they differ from the register implicit machine only in where the operands are located. As an example, the addition instruction starts in phase Ad_2 by fetching its first operand from the first source register into the DR register, then in phase Ad_3 the second operand is fetched from the second source register into the SR register, and finally, in Ad_4 the result is written to the destination register.

The last set of instructions is the set of branch instructions, both conditional and unconditional. These instructions remain the same on the register machine, as they were on the register implicit machine.

7.2.2   Control for the Accumulator Machine

The accumulator architecture is a little further from the register implicit architecture than the register architecture. Although this is true, the differences are manageable, and the RIM machine can be implemented for an accumulator architecture with relative ease.

There are a couple of significant differences between the register implicit machine and the accumulator machine. Maybe the most noticeable difference is the presence of the AC register in the accumulator machine. However, this register essentially serves the same purpose as the source register, SR, in the register implicit machine, and so this change really only requires that we rename the SR register as AC.

The second noticeable difference in the two machines is the use of implicit mode operands in the accumulator machine, a mode which is not used in the register implicit machine. In the accumulator machine, frequently the AC register is an implicit operand in an operation.

The previous paragraphs expose something rather interesting about the accumulator machine; what seemed like a significantly different machine from the register implicit machine, turns out to be a machine that is very similar. As a matter of fact, the data-paths of the two machines are identical, except for renaming the SR register as the AC register. It turns out that most of the difference between the two machines is found in the control unit.

The control specification for the accumulator machine is given in Table 7.7. The fetch phases are identical to those of the register implicit machine. The load instruction is identical to the load instruction of the register machine, without the last phase, L_4, where the operand is put in the register file. For the store instruction, the operand comes from the AC register, rather than the register file. This transfer is done in phase St_3, and requires manipulation of the ALU, since anything coming from the AC onto the bus, must pass through the ALU, with an identity operation.

The ALU operations are executed in two phases. For example, for the addition instruction, in phase Ad_2 the explicit operand is loaded into the DR register. In phase Ad_3, the DR register is added to the AC, and the Z flag is set, according to the result.

For the branch instructions, the phases for the accumulator machine are identical to those of the register implicit machine.

Notice that the SRF control line and MUX are never used in the Accumulator machine. Option 0 is always used to address the register file, and so, the MUX could be removed from the data-path, and the address line of the register file could be directly connected to IR_{6-4}.

TABLE 7.7   Micro-code for the accumulator machine.

Phase	Micro-instruction	Control Output
F_0	$T_0 : AR \leftarrow PC, C \leftarrow C + 1$	ARLD, SPC, CIN
F_1	$T_1 : IR \leftarrow M[AR], PC \leftarrow PC + 1,$ $C \leftarrow C + 1$	IRLD, ME, PCIN, CIN
L_2	$T_2 \cdot OP_0 : AR \leftarrow IR_{7-0}, C \leftarrow C + 1$	ARLD, SAD, CIN
L_3	$T_3 \cdot OP_0 : AC \leftarrow M[AR],$ $C \leftarrow 0$	ACLD, ME, CCL
St_2	$T_2 \cdot OP_1 : AR \leftarrow IR_{7-0}, C \leftarrow C + 1$	ARLD, SAD, CIN
St_3	$T_3 \cdot OP_1 : M[AR] \leftarrow AC, C \leftarrow 0$	MW, SALU, ALUOP=100, CCL
Ad_2	$T_2 \cdot OP_2 : DR \leftarrow R[IR_{2-0}], C \leftarrow C + 1$	DRLD, RE, SS, CIN
Ad_3	$T_3 \cdot OP_2 : AC \leftarrow AC + DR,$ $Z \leftarrow (AC + DR) = 0, C \leftarrow 0$	ACLD, SALU, ZLD, CCL, ALUOP = 000
Sb_2	$T_2 \cdot OP_3 : DR \leftarrow R[IR_{2-0}], C \leftarrow C + 1$	DRLD, RE, SS, CIN
Sb_3	$T_3 \cdot OP_3 : AC \leftarrow AC - DR,$ $Z \leftarrow (AC - DR) = 0, C \leftarrow 0$	ACLD, SALU, ZLD, CCL, ALUOP = 011
An_2	$T_2 \cdot OP_4 : DR \leftarrow R[IR_{2-0}], C \leftarrow C + 1$	DRLD, RE, SS, CIN
An_3	$T_3 \cdot OP_4 : AC \leftarrow AC \wedge DR,$ $Z \leftarrow (AC \wedge DR) = 0, C \leftarrow 0$	ACLD, SALU, ZLD, CCL, ALUOP = 111
Or_2	$T_2 \cdot OP_5 : DR \leftarrow R[IR_{2-0}], C \leftarrow C + 1$	DRLD, RE, SS, CIN
Or_3	$T_3 \cdot OP_5 : AC \leftarrow AC \vee DR,$ $Z \leftarrow (AC \vee DR) = 0, C \leftarrow 0$	ACLD, SALU, ZLD, CCL, ALUOP = 110
N_2	$T_2 \cdot OP_6 : DR \leftarrow R[IR_{2-0}], C \leftarrow C + 1$	DRLD, RE, SS, CIN
N_3	$T_3 \cdot OP_6 : AC \leftarrow \overline{AC},$ $Z \leftarrow \overline{AC} = 0, C \leftarrow 0$	ACLD, SALU, ZLD, CCL, ALUOP = 101
Bz_2	$T_2 \cdot OP_8 \cdot ZX : PC \leftarrow IR_{7-0}$	PCLD, SAD
$Bz_2 \vee \overline{Bz_2}$	$T_2 \cdot OP_8 : C \leftarrow 0$	CCL
Bn_2	$T_2 \cdot OP_9 \cdot \overline{ZX} : PC \leftarrow IR_{7-0}$	PCLD, SAD
$Bn_2 \vee \overline{Bn_2}$	$T_2 \cdot OP_9 : C \leftarrow 0$	CCL
J_2	$T_2 \cdot OP_7 : PC \leftarrow IR_{7-0}, C \leftarrow 0$	PCLD, SAD, CCL

7.2.3   Control for the Stack Machine

The final architecture we consider is the stack machine. There is a major difference between this architecture, and the other three architectures. Rather than accessing the register file by specifying an address, we now access it using the operations *pop* and *push*. So, we must set the register file up as a stack.

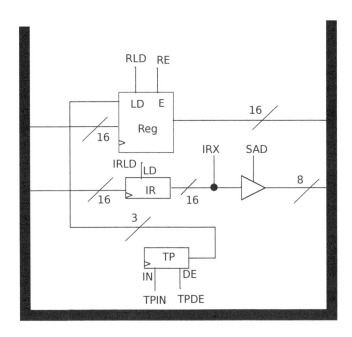

FIGURE 7.13 Stack machine data-path modifications.

In Figure 7.13 we show the required changes to the BRIM data-path, for the stack machine. Addressing the register file has been dramatically simplified. The address always comes from the TP register (top pointer). This new register is used to define a stack, using the common *array, and top pointer* implementation.

We need to be able to pop and push the stack, and so, instead of our standard register, we are using an up-down counter to implement the TP register. The up-down counter can be incremented, using the IN control input, or decremented, using the DE control input. In our stack implementation, $R[7]$ will be the base of the stack. A *pop* operation will increment the TP register, and a *push* operation will decrement the TP register.

TABLE 7.8   Micro-code for the stack machine.

Phase	Micro-instruction	Control Output
F_0	$T_0 : AR \leftarrow PC, C \leftarrow C+1$	ARLD, SPC, CIN
F_1	$T_1 : IR \leftarrow M[AR], PC \leftarrow PC+1,$ $C \leftarrow C+1$	IRLD, ME, PCIN, CIN
Pu_2	$T_2 \cdot OP_0 : AR \leftarrow IR_{7-0},$ $TP \leftarrow TP-1, C \leftarrow C+1$	ARLD, SAD, TPDE, CIN
PuM_3	$T_3 \cdot OP_0 \cdot \overline{IRX_{11}} : R[TP] \leftarrow M[AR]$	RLD, ME
PuI_3	$T_3 \cdot OP_0 \cdot IRX_{11} : R[TP] \leftarrow IR_{7-0}$	RLD, SAD
$PuM_3 \vee$ PuI_3	$T_3 \cdot OP_0 : C \leftarrow 0$	CCL
Po_2	$T_2 \cdot OP_1 : AR \leftarrow IR_{7-0}, C \leftarrow C+1$	ARLD, SAD, CIN
Po_3	$T_3 \cdot OP_1 : M[AR] \leftarrow R[TP],$ $TP \leftarrow TP+1, C \leftarrow 0$	MW, RE, TPIN, CCL
Ad_2	$T_2 \cdot OP_2 : DR \leftarrow R[TP],$ $TP \leftarrow TP+1, C \leftarrow C+1$	DRLD, RE, TPIN, CIN
Ad_3	$T_3 \cdot OP_2 : SR \leftarrow R[TP], C \leftarrow C+1$	SRLD, RE, CIN
Ad_4	$T_4 \cdot OP_2 : R[TP] \leftarrow SR + DR,$ $Z \leftarrow (DR + SR) = 0, C \leftarrow 0$	RW, SALU, ZLD, CCL, ALUOP = 000
Sb_2	$T_2 \cdot OP_3 : DR \leftarrow R[TP],$ $TP \leftarrow TP+1, C \leftarrow C+1$	DRLD, RE, TPIN, CIN
Sb_3	$T_3 \cdot OP_3 : SR \leftarrow R[TP], C \leftarrow C+1$	SRLD, RE, CIN
Sb_4	$T_4 \cdot OP_3 : R[TP] \leftarrow SR - DR,$ $Z \leftarrow (DR - SR) = 0, C \leftarrow 0$	RW, SALU, ZLD, CCL, ALUOP = 011
An_2	$T_2 \cdot OP_4 : DR \leftarrow R[TP],$ $TP \leftarrow TP+1, C \leftarrow C+1$	DRLD, RE, TPIN, CIN
An_3	$T_3 \cdot OP_4 : SR \leftarrow R[TP], C \leftarrow C+1$	SRLD, RE, CIN
An_4	$T_4 \cdot OP_4 : R[TP] \leftarrow SR \wedge DR,$ $Z \leftarrow (DR \wedge SR) = 0, C \leftarrow 0$	RW, SALU, ZLD, CCL, ALUOP = 111
Or_2	$T_2 \cdot OP_5 : DR \leftarrow R[TP],$ $TP \leftarrow TP+1, C \leftarrow C+1$	DRLD, RE, TPIN, CIN
Or_3	$T_3 \cdot OP_5 : SR \leftarrow R[TP], C \leftarrow C+1$	SRLD, RE, CIN
Or_4	$T_4 \cdot OP_5 : R[TP] \leftarrow SR \vee DR,$ $Z \leftarrow (DR \vee SR) = 0, C \leftarrow 0$	RW, SALU, ZLD, CCL, ALUOP = 110
N_2	$T_2 \cdot OP_6 : DR \leftarrow R[TP], C \leftarrow C+1$	DRLD, RE, CIN
N_3	$T_3 \cdot OP_6 : R[TP] \leftarrow \overline{DR},$ $Z \leftarrow (\overline{DR}) = 0, C \leftarrow 0$	RW, SALU, ZLD, CCL, ALUOP = 101
Bz_2	$T_2 \cdot OP_8 \cdot ZX : PC \leftarrow R[TP],$ $TP \leftarrow TP+1$	PCLD, RE, TPIN
$Bz_2 \vee$ $\overline{Bz_2}$	$T_2 \cdot OP_8 : C \leftarrow 0$	CCL
Bn_2	$T_2 \cdot OP_9 \cdot \overline{ZX} : PC \leftarrow R[TP],$ $TP \leftarrow TP+1$	PCLD, RE, TPIN
$Bn_2 \vee$ $\overline{Bn_2}$	$T_2 \cdot OP_9 : C \leftarrow 0$	CCL
J_2	$T_2 \cdot OP_7 : PC \leftarrow R[TP],$ $TP \leftarrow TP+1, C \leftarrow 0$	PCLD, RE, CCL

Control specifications for the stack machine are presented in Table 7.8. We see that the specifications for the fetch phases, F_0 and F_1 have not changed from the previously considered architectures. For the stack machine, the $movR/M$ instruction has been replaced by the push instruction, and the store instruction has been replaced by the pop instruction.

Examining the push instruction, execution starts in phase Pu_2, and then continues to either phase PuM_3 or PuI_3, depending on whether direct or immediate mode is being used. In phase Pu_2, we load the address field of the instruction into the AR register, preparing for a potential memory read, and decrement the TP register, allocating a new register to the top of the stack, to receive the value being pushed. In phase PuM_3 we complete the execution for a memory direct operand, by reading the operand from memory, and placing it in the register file, at the newly allocated top spot. Alternatively, in phase PuI_3 we complete the execution for an immediate mode operand, by taking the immediate value from the instruction, and loading it into the new top slot in the stack. Regardless of whether phase PuM_3 or phase PuI_3 is performed, the table line marked $PuM_3 \vee PuI_3$ indicates that at time step T_3 the sequencer is reset, causing the machine cycle to start over.

For a pop instruction we perform the actions of phases Po_2 and Po_3. This sequence should be relatively straight-forward to understand. In phase Po_2 the address is loaded into the AR register. In phase Po_3 the operand is read from the top of the stack, and written to the memory unit, and the TP register is incremented, popping the operand off the top of the stack.

Next, we move on to the ALU instructions. Since these instructions are all very similar, we only consider the addition instruction, that is executed by performing the three phases, Ad_2, Ad_3, and Ad_4. Starting in phase Ad_2, the top element of the arithmetic stack is placed into the DR register, and this element is popped off of the stack by incrementing the TP register. In phase Ad_3, the next element on the stack is placed into the SR register. This time the stack is, however, not popped. Instead, in phase Ad_4, the result of the addition is put in the unpopped slot. The effect is to pop two operands, and push the result. Also, in phase Ad_4, the Z flag is updated to reflect the status of the result.

For the branch instructions, the structure of the micro-code is similar to that for the other architectures. The difference is where the jump address is coming from. In previous architectures, the jump address was always an immediate value, coming from the instruction. For the stack machine, the jump address is assumed to be on the stack. So, for example, in phase J_2 the PC is loaded from the top of the arithmetic stack, and the arithmetic stack

TABLE 7.9   Machine code for Exercise 7.1.

Assembly Code	Machine Code	Meaning
dbl R	1110 0 R 00000000	$R \leftarrow 2 \times R_1$

is simultaneously popped. You can compare this with the other architectures, in which the PC is loaded from the IR register.

7.3   SUMMARY

We have now given specifications for the processors of all four architectures which we discussed. These specifications consist of a data-path design and a behavioral description of the control unit. From these specifications, it is possible to implement these processors, physically. This could be done by translating our descriptions into, say, Verilog, and by using appropriate authoring tools, we could easily program the combinational circuitry into a PROM unit, and use a SRAM unit for state storage.[1]

The RIM computer, which we have constructed is, of course, a fairly simplistic computer. In Chapter 6, we point out several of its shortcomings. For example, it has no hardware support for floating-point arithmetic. But, it is interesting that there are many processors that are capable of performing floating-point arithmetic, but that have no floating-point hardware. Instead, floating-point is handled by a software library. So that, although the RIM processor is simplistic, it is, in fact, capable of doing most of the operations you can do on your favorite general-purpose processor.

The technique we have used to construct the RIM processor is no more than an application of the techniques described in Chapter 5. The processor is an algorithmic circuit. We converted the algorithm that the processor executes, the machine cycle, into an RTL description, by first drawing a flow chart and data path, and then translating the flowchart into RTL. From the RTL we can further use the techniques of Chapter 5, to implement the circuitry.

7.4   EXERCISES

7.1   We wish to add the machine instruction *dbl* to the register implicit machine. This instruction doubles the value in a given register. Its specification is given in Table 7.9. Make modifications to the BRIM machine

---

[1] An easier option is to use a Field Programmable Gate Array (FPGA). An FPGA is a device that contains both memory components, and gate-level logic components, that are unconnected. These components can be connected into custom circuitry, in a process called *programming*. FPGA units are also re-programmable, allowing easy design experimentation.

TABLE 7.10   Machine code for Exercise 7.2.

Assembly Code	Machine Instruction Code	Meaning
skip	1110 0 000 00000000	$PC \leftarrow PC + 1$

TABLE 7.11   Machine code for Exercise 7.3.

Assembly Code	Machine Code	Meaning
swap R_1, R_2	1110 0 R_1 0 R_2 0000	$R[R_1] \leftarrow R[R_2], R[R_2] \leftarrow R[R_1]$

to include this instruction. In particular, make the necessary changes to Figure 7.2, Figure 7.11, and Tables 7.2 and 7.3.

7.2   This exercise is similar to Exercise 7.1. Instead, you are adding the instruction *skip* that skips the next instruction. The specifications are given in Table 7.10.

7.3   This exercise is similar to Exercise 7.1. Instead, you are adding the instruction *swap* that swaps the contents of two registers. The specifications are given in Table 7.11. (Assume that the addressing mode for R_2 is always direct mode.)

7.4   In Tables 7.2 and 7.3, when we presented the BRIM machine we did not include many useful addressing modes, like the indirect mode. In this exercise, we add the indirect mode to the BRIM machine. Make the figure and table modifications requested in Exercise 7.1 to add the instruction *storei* to the BRIM machine. It is defined in Table 7.12.

7.5   Design a machine with the following specification.

- A word size of 8 bits.

- A RAM unit of size 64×8.

- Two machine registers R and S.

- A two-operation ALU: an addition operation (ALUOP $= 0$), and an OR operation (ALUOP $= 1$).

- An ISA described in Table 7.13.

Develop design tools like Figure 7.2, Figure 7.11, and Tables 7.2 and 7.3.

TABLE 7.12   Machine code for Exercise 7.4.

Assembly Code	Machine Code	Meaning
storei m, R_1	1110 0 R_1 m	$M[M[m]] \leftarrow R[R_1]$

TABLE 7.13   ISA for Exercise 7.5.

Assembly Code	Machine Code	Meaning
add	00 000000	$R \leftarrow R + S$
store m	01 m	$M[m] \leftarrow R$
or	10 000000	$S \leftarrow R \vee S$
jump m	11 m	$PC \leftarrow m$

TABLE 7.14   Micro-code for Exercise 7.6.

Phase	Micro-instruction	Control Output
F_0	$T_0 : AR \leftarrow PC, C \leftarrow C + 1$	ARLD, SPC, CIN
F_1	$T_1 : IR \leftarrow M[AR], PC \leftarrow PC + 1,$ $C \leftarrow C + 1$	IRLD, ME, PCIN, CIN
$I0_2$	$T_2 \cdot OP_0 : PC \leftarrow PC + 1, C \leftarrow 0$	PCIN, CCL
$I1_2$	$T_2 \cdot OP_1 : R \leftarrow R \wedge S, C \leftarrow 0$	RLD, SALU, CCL
$I2_2$	$T_2 \cdot OP_2 : AR \leftarrow IR_{5-0}, C \leftarrow C + 1$	ARLD, SAD, CIN
$I2_3$	$T_3 \cdot OP_2 : S \leftarrow M[AR], C \leftarrow 0$	SLD, ME, CCL
$I3_2$	$T_2 \cdot OP_3 : R \leftarrow R - S, C \leftarrow 0$	RLD, ALUOP, SALU, CCL

7.6   You are working with a machine with a data-path described in Exercise 7.5. The ISA is different. Table 7.14 gives the control specification for the machine. Draw the data-path for the machine, with a diagram similar to Figure 7.2. Also develop a description of the ISA for the machine, similar to Table 6.1. Give the instructions meaningful assembly language names.

7.7   Build the control unit for the machine described in Exercise 7.6. First you will need to develop the equations for the outputs, and then you will be able to draw circuitry from the equations.

7.8   Design an ALU that performs the following operations. Draw a diagram of the ALU, similar to Figure 7.6.

• Increment: $Z \leftarrow A + 1$.

• Negative: $Z \leftarrow -A$.

• NOR: $Z \leftarrow \overline{A \vee B}$.

• Identity: $Z \leftarrow A$.

• XOR: $Z \leftarrow A \oplus B$.

7.9   We wish to add the machine instruction *clear* to the register machine.

TABLE 7.15   Machine code for Exercise 7.9.

Assembly Code	Machine Code	Meaning
clear R	1110 0 R 00000000	$R \leftarrow 0$

TABLE 7.16   Machine code for Exercise 7.10.

Assembly Code	Machine Code	Meaning
clear	1110 0 000 00000000	$AC \leftarrow 0$

TABLE 7.17   Machine code for Exercise 7.11.

Assembly Code	Machine Code	Meaning
dup	1110 000000000000	$push_1(pop), push_2(pop)$

It resets a particular register to zero. This instruction can be performed using the ALU. The specification for the clear instruction is given in Table 7.15. Make modifications to the register machine to accommodate this new instruction. In particular, make modifications to Figure 7.2, for the data-path, and Tables 7.5 and 7.6, for the control unit.

7.10 This exercise is similar to Exercise 7.9. Instead of the register machine, implement the *clear* instruction on the accumulator machine. Table 7.16 gives the semantics of the new instruction. Make your modifications to Table 7.7.

7.11 We wish to add the machine instruction *dup* to the stack machine. It duplicates the top element of the stack, leaving the stack with two copies of this element on the top. The specification for the duplicate instruction is given in Table 7.17, where $push_1$ and $push_2$ represent the first and second push, respectively. Make modifications to the stack machine to accommodate this new instruction. In particular, make modifications to Table 7.8.

7.12 Develop the output equations for the following control lines of the stack machine used in Table 7.8.

   a. SRLD

   b. TPIN

   c. CIN

   d. $ALUOP_0$

Computer Arithmetic

CONTENTS

IN CHAPTER 7, we presented a fairly simple ALU. As a matter of fact, we presented two arithmetic-logic units: an extremely simple two-operation unit, and the six-operation ALU used in the BRIM machine. But, even the BRIM ALU had a limited set of operations.

A fuller name for the ALU might be the ALSU, or Arithmetic Logic Shift Unit. In this chapter, we explore what a more typical ALSU would be like. To do this, we must examine the more common arithmetic, logic, and shift operations.

Remember that an ALSU is, essentially, an output multiplexer, connected to computational units; each unit computing one of the operations of the ALSU. The major question which needs to be answered is what do these computational units look like. For the logic and shift operations, the answer is fairly straightforward, and so most of what we concentrate on, in this chapter, is arithmetic. However, before we launch into arithmetic, it is worth spending a little time talking about the logic and shift units.

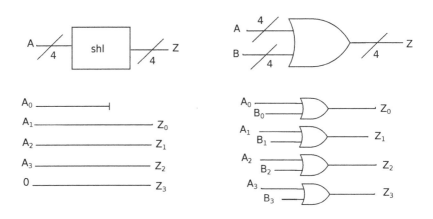

FIGURE 8.1 *Shr* and OR computational units.

8.1 LOGIC AND SHIFT OPERATIONS

We can illustrate the operation of a logic computational unit and a shift computational unit, with a couple of examples. In Figure 8.1, we show a logical shift-right unit, and an OR unit for a 4-bit machine. As you can see, the shift unit is nothing more than a re-routing of input lines, to output lines. The OR unit is an array of OR gates, each computing the result of one of the output bit positions. These computational units can be widened by increasing the array size.

It should be easy to see, from Figure 8.1, how the other logic and shift functions are implemented. For example, an AND computational unit is constructed in the same way as the OR unit, substituting AND gates for OR gates. And, a circular left-shift unit is constructed in a similar fashion as the logical shift unit, but routing the high-order bit of the input down to the low-order bit of the output.

8.2 ARITHMETIC OPERATIONS

We now turn our attention to arithmetic. Arithmetic operations are performed on what we might call *numeric data*. Numeric data are bit collections which we view as representing numbers of some kind. To do justice to our discussion of arithmetic we need to understand the different types of numeric data, in more detail. Although there are other types of numeric data, there are two types that are almost universally supported by general-purpose processors.

- Integer data.

- Floating-point data.

We have been using integer data throughout the chapters of this book, and, on several occasions, we have mentioned floating-point data. In the next few sections we examine the arithmetic performed on these two data-types, beginning with integer data.

8.2.1 Unsigned and Signed Integers

As discussed in Section 2.1.3, most computers implement at least two integer data-types.

- Unsigned integer.

- Signed integer.

In the unsigned integer type, only non-negative integers are represented. What we mean by this is that all bit configurations of the computer word are used to represent integer values greater than or equal to zero. In the signed integer type, both positive and negative numbers are represented. So, some of the possible bit configurations are used to represent negative integers, and some are used to represent non-negative integers.

The reason that these two integer types exist is that computers deal with integers in, at least, two different ways. Often the computer is working with memory addresses, which are always non-negative. For this use, unsigned integers, which use all bit configurations to represent non-negative integers, allow for the representation of a larger range of addresses. On the other hand, often computers are doing arithmetic computations on general integer values. For this use, negative values are needed, and the signed integer type is necessary.

Let us talk about what we mean by implementing the two types of integers. Suppose that we are working on an 8-bit machine. And, we wish to perform the operation $R0 \leftarrow R0 + R1$. Further, assume that R0 = 00001001, and that R1 = 11110011. The bit configurations given for R0 and R1, are just that: bit configurations. The numbers these represent are determined by their interpretation. If we interpret them as unsigned integers, they are both positive, and the result we obtain from the addition operation is also positive. If we interpret them as signed integers, R0, with a 0 as its sign bit, is positive, and R1, with a 1 as its sign bit, is negative, and the result we obtain from the addition, with 1 as its sign bit, is negative.

There is nothing in the bit configurations for R0 and R1 that tells the processor whether they should be treated as signed or unsigned. What tells the processor how to treat these registers is the instruction that is used to do the addition. A computer supporting both signed and unsigned arithmetic, would probably have two different addition instructions. The following instruction might be used to add the register values, treating them as unsigned values.

```
addu R0, R1
```

Alternatively, the following instruction might be used to add them as signed integer values.

```
addi R0, R1
```

So, the processor would support two distinct, but matching sets of arithmetic instructions: one for unsigned arithmetic, and one for signed arithmetic.

Let us now examine the characteristics of our representations for signed and unsigned integers. We have covered how numbers are stored in unsigned representation in Chapter 2. As an example, if we are working on an 8-bit machine, if we wish to represent a non-negative number, like 57, in the computer word, we simply convert this number into binary, and extend it out to eight bits with leading zeros. So, 57, in decimal, is 111001 in binary, which gives us the 8-bit representation of 00111001.

In our 8-bit unsigned system, the smallest number we can represent is 00000000, or zero. The largest number is 11111111, which, if you convert it to decimal is $255 = 2^8 - 1$. In general, if you have a k-bit word, the largest number you can represent, as an unsigned integer, is $2^k - 1$. The important point, in this discussion, is that we cannot represent all possible integers; only integers that fall within a certain finite range.

Turning to signed integers, remember from Section 2.1.3, that in our two's compliment representation, the high-order bit of the word is used as the sign bit; a 0 in the sign bit indicates a non-negative number, and a 1 in the sign bit indicates a negative number. If we count through the signed integers, in the positive direction, starting at 0, we get the sequence of bit configurations 00000000, 00000001, 00000010, ..., 01111111. This represents the sequence of integers $0, 1, 2, \ldots, 127 = 2^7 - 1$. Counting in the negative direction, starting at zero, we get the sequence 00000000, 11111111, 11111110, 11111101, 11111100, ..., 10000000. You may not notice it, but what we are doing is just counting in binary, using the 1 instead of the 0, and the 0 instead of the 1. In any case, we are counting the sequence $0, -1, -2, -3, -4, \ldots, -2^7$. So, the range of our 8-bit signed integers is $2^7 - 1$ to -2^7. As with the unsigned representation, for the signed representation we also have a finite range. In general, with a k-bit word, we have signed integers from $2^{k-1} - 1$ to -2^{k-1}. You will notice that there is one more negative number than non-negative numbers. This is because 0 takes up one of the non-negative bit configurations, whereas all negative bit configurations are used for counting negative numbers.

8.2.2 Unsigned Arithmetic

Once we understand the basic characteristics of integer representation, we can begin discussing computational units that compute standard arithmetic operations. We would expect any modern computer to be able to perform the usual multiplicative and additive operations.

- $Z \leftarrow A + B$

- $Z \leftarrow A - B$

- $Z \leftarrow A \times B$

- $Z \leftarrow A \div B$

- $Z \leftarrow A \bmod B$

These operations would need to be implemented for both signed and unsigned integers. We begin by discussing arithmetic for unsigned integers.

8.2.2.1 Unsigned Addition

The computational unit, used to perform unsigned integer addition, is the ripple-carry adder, covered in Section 3.1.2.6. To perform the addition $Z \leftarrow A + B$, we would input the values A and B into the adder, together with a carry-in of 0. The output of the adder would then be the result, Z. Remember that the adder also produces a carry-out, as a result. This carry-out is typically "discarded." When we say "discarded," we mean that it is not part of the result Z. However it does have another role.

To explain the other role of the carry-out output, we present a few addition examples. We start by showing the computation of $00110100 + 10001101$.

$$
\begin{array}{r}
{\scriptstyle 001111000} \\
00110100 \\
+10001101 \\
\hline
11000001
\end{array}
\qquad (8.1)
$$

In decimal, Computation 8.1 represents $52 + 141 = 193$. Now let's look at the computation $10010111 + 01110100$.

$$
\begin{array}{r}
{\scriptstyle 111101000} \\
10010111 \\
+01110100 \\
\hline
00001011
\end{array}
\qquad (8.2)
$$

You should sense that there is something very wrong with Computation 8.2. In particular, it appears, and correctly so, that the result is smaller than the two addends. In fact, in decimal, Computation 8.2 computes $151 + 116 = 11$, which is clearly incorrect. The correct result is 267.

We have just observed a phenomenon called *overflow*. The problem is that we had a 1 as the carry-out. If we discard the carry-out, as usual, we lose the most significant bit of the result. And, the root problem that has caused this overflow, is that we have computed an additive result that is too large to be represented with just eight bits. The largest unsigned value that can be represented with eight bits is 255, or $2^8 - 1$.

So how does the processor handle an arithmetic overflow? The short answer to this question is that it doesn't, usually. The best it can normally do is notify the user that an overflow has occurred, and let the user handle the condition. The user is notified of an overflow by a status flag, which we will call V. Remember, from Chapter 7, that the ALU already produces the Z status flag, indicating that the result of a computation is zero. The Z flag is not the only status flag produced by the ALU. Another flag would be the V flag.

The next question, that comes to mind, is how the V flag is computed. You may already have figured this out, by observing the two computations 8.1 and 8.2. Computation 8.1 is an example in which there is no overflow. Notice that, equally, there is no carry-out. In Computation 8.2, there is overflow, and equally, there is a carry-out. This brings us to the Boolean equation for the flag V, when performing unsigned addition.

$$
V = C_{out}
\qquad (8.3)
$$

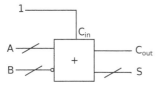

FIGURE 8.2 A subtraction circuit.

8.2.2.2 Unsigned Subtraction

We like to say that computers do not perform subtraction; they only perform addition. Although computers, obviously, do perform subtraction, this
statement is meant to indicate that processors contain very little circuitry
dedicated to subtraction, but rather use the same circuitry they use for addition. The way the circuitry is reused can be seen, if we rewrite the subtraction
operation $Z \leftarrow A - B$ as $Z \leftarrow A + (-B)$. Specifically, for the computer to
subtract B from A, we can simply add the two's compliment of B to A. The
circuit that performs this is shown in Figure 8.2. An adder is used to add the
operand A, the one's compliment of B, and a carry-in of one. The carry-in
and the one's compliment form the two's compliment of B.

Since subtraction is performed by doing addition, we also have the potential
for overflow. Let's do a couple of examples to illustrate the nature of overflow
in subtraction. We start with the computation $00110111 - 00001110$.

$$
\begin{array}{r}
{\scriptstyle 111101111} \\
00110111 \\
+11110001 \\
\hline
00101001
\end{array}
\qquad (8.4)
$$

Computation 8.4 represents the decimal computation $55 - 14 = 41$ We have
written the subtraction operation as the minuend, 00110111, added to the
one's compliment of the subtrahend, with a carry-in to the low-order bit of
1. (This forms the two's compliment of the subtrahend; its one's compliment
plus 1.)

Notice that Computation 8.4 has a carry-out. For addition, this would indicate an overflow condition, and yet, here we see that we got the correct result,
and so there was no overflow. This would imply that there is a different overflow criterion for subtraction, than for addition.

The obvious question is then, what is the criterion used for overflow calculation in subtraction? To develop this criterion, let us do an example that

we know will result in overflow: $00111001 - 01111000$. The reason we know this will result in overflow becomes apparent when we convert this problem into decimal: $57 - 112 = -55$. Since we cannot represent negative numbers in unsigned notation, it is clear that any result we get will be in error.

$$
\begin{array}{r}
\scriptstyle 011111111 \\
00111001 \\
+10000111 \\
\hline
11000001
\end{array}
\tag{8.5}
$$

We see that the result we obtain from Computation 8.5 is garbage, as expected; it is larger that both of the operands, which should not happen with subtraction. But, what is more interesting is what happened to the carry-out. Computation 8.5 does not produce a carry-out. And so, we see that for unsigned subtraction the calculation of the overflow flag V is reversed from how it is calculated for addition. That is to say, we have no overflow condition when there is a carry, and we have overflow when there is no carry:

$$
V = \overline{C_{out}}
\tag{8.6}
$$

8.2.2.3 Unsigned Multiplication

For addition, we were already familiar with the circuit used by the ALSU as the computational unit: the adder. For multiplication, we must construct a multiplier circuit to perform the operation $Z \leftarrow A \times B$. But, before we can do that, we must understand how multiplication is performed in binary.

We start by presenting a method for binary multiplication that we humans might use. Then we transform this method into a method that we can implement in circuitry.

To illustrate how a human might multiply two numbers in binary, we present an example in four bits: 0110×1011. Notice that in decimal, this computation is $6 \times 11 = 66$. Our multiplication method is just to apply what we call long-multiplication to binary numbers.

$$
\begin{array}{r}
0110 \\
\times 1011 \\
\hline
0110 \\
0110 \\
0000 \\
0110 \\
\hline
1000010
\end{array}
\tag{8.7}
$$

In Computation 8.7, just as in decimal, we write several intermediate rows, before summing them to produce the product. Each row is one of the digits of the multiplier, multiplied by the multiplicand. In a binary calculation, since a digit of the multiplier is ether a 0 or a 1, each row is, simply, a row of zeros, or a copy of the multiplicand, respectively.

We now look at how this long-multiplication method can be adapted to work with circuitry. To implement this method in circuitry, we will need to store operands, and a result, in registers. In this discussion, we will call the registers N, M, and P, for the multiplicand, multiplier, and product, respectively. Let's assume that we are performing 4-bit multiplication. And so, the question we next ask is how wide are these three registers? The multiplier is a 4-bit value, and so, we will make the M register four bits. The product register, P, needs to be eight bits wide. This register must be able to store the largest product that can be produced by multiplying two 4-bit numbers; this is $1111 \times 1111 = 11100001$, which has an 8-bit product. Finally, the multiplicand register, N, must be eight bits wide also. The reason for this is that it is added to the product register, P, which is an 8-bit register, and our ripple-carry adder is only capable of adding two numbers of the same width. To summarize, our circuit has the following arithmetic registers.

- The N register: 8 bits wide. For k-bit multiplication, this register would be $2k$ bits wide.

- The M register: 4 bits wide. For k-bit multiplication, this register would be k bits wide.

- The P register: 8 bits wide. For k-bit arithmetic, this register would be $2k$ bits wide.

We are going to make a few modifications to the usual long-multiplication method, to make it better conditioned for implementation in circuitry. If we examine the long-multiplication method, we see that we write down a sequence of rows. Each row is either a copy of the multiplicand, shifted to the left, or a row of zeros. Each row represents a value that is added into the product at the end of the procedure to produce the result. The first modification we make is that we do not wait until the end of the procedure to add the rows. Rather, we add each row to the product, P, as it is calculated.

Now, so far what we have is that the multiplicand is shifted left, and added to the product, which is held stationary. Our second modification is to change this shifting scheme. Instead of shifting the multiplicand to the left, what we do is shift the product right, and hold the multiplicand stationary. This has the same effect as shifting the multiplicand left, and holding the product stationary.

TABLE 8.1   Trace of Multiplication 8.7.

N	M	P	Action
01100000	1011	00000000	$+$
		01100000	$\rightarrow$
	0101	00110000	$+$
		10010000	$\rightarrow$
	0010	01001000	$\rightarrow$
	0001	00100100	$+$
		10000100	$\rightarrow$
	0000	01000010	$\checkmark$

With these modifications, let's look at how we would perform Calculation 8.7 in the registers N, M, and P. Table 8.1 shows a trace of the modified long-multiplication method. Initially the N register starts with the value of the multiplicand, shifted into the upper half. The register M is initialized to the multiplier, and the register P is set to zero. The algorithm then makes four passes through a loop. (For k-bit multiplication, you would make k passes.) On each pass the multiplier, M, and the product register, P, are shifted one bit to the right. In addition, on each pass the multiplicand might, or might not be added to the product. If the low-order bit of the multiplier, M_0, is 1, the addition is performed. If M_0 is 0, then the addition is skipped.

Table 8.1 shows the value of N in the first column. The contents of this register remain the same throughout the algorithm. The second, and third columns show the values of the M and P registers, as the algorithm progresses. The last column shows the action performed: either an addition, a shift-right, or in the final row, the algorithm stops. In the first row, M_0 is a 1, so the N register is added to the P register. Notice that because the low-order bits of the N register are zeros, the addition only changes the high-order bits of the P register. In the second row, the M and P registers are shifted right, completing the first pass through a loop. In the third row, M_0 is 1 again, so another addition is performed. This is followed by a shift, in the fourth row, to complete the second pass. In the fifth row, M_0 is a 0, so only a shift is performed, to complete the third pass. In the sixth row, M_0 is a 1 again, so that an addition is performed, followed by a shift in the seventh row, to complete the forth pass. In the eighth row, we show the final result, after the algorithm has stopped.

The algorithm that we used in Table 8.1 is given in pseudo-code below. It starts by shifting the N register, so that the operand is in the leftmost four bits, and setting the P register to zero. A register, called I, is used to count the number of times the loop is executed. In the loop, we simply check to see if the addition is required, and if so perform it, and then shift the M and P registers. Notice that we use the notation $X >> 1$ to indicate a shift right of X by one bit. In a similar way, the notation $X << 1$ indicates a 1-bit left shift.

$$
\begin{aligned}
&N_{7-4} = N_{3-0} \\
&N_{3-0} = 0 \\
&P = 0 \\
&I = 0 \\
&\texttt{while } I \neq 4 \texttt{ do} \\
&\qquad \texttt{if } M_0 \texttt{ then} \\
&\qquad\qquad P = P + N \\
&\qquad M = M >> 1 \\
&\qquad P = P >> 1 \\
&\qquad I = I + 1
\end{aligned}
\tag{8.8}
$$

We can turn this algorithm into circuitry, using the method explained in Section 5.3. We give the behavioral description of the circuit below.

$$
\begin{aligned}
&\text{Def: } I4X \equiv I = 4, T_k \equiv C = k \\
&T_0 : N_{7-4} \leftarrow N_{3-0}, N_{3-0} \leftarrow 0, P \leftarrow 0, I \leftarrow 0, C \leftarrow C + 1 \\
&T_1 \cdot I4X : C \leftarrow 4 \\
&T_1 \cdot \overline{I4X} : C \leftarrow 2 \\
&T_2 \cdot M_0 : P \leftarrow P + N \\
&T_2 : C \leftarrow C + 1 \\
&T_3 : M \leftarrow shr\, M, P \leftarrow shr\, P, I \leftarrow I - 1, C \leftarrow 1 \\
&T_4 :
\end{aligned}
\tag{8.9}
$$

This description starts with a definition of a signal $I4X$ that, if 1, indicates that $I = 4$. Time step T_0 does the initialization. Time step T_1 checks to see if $I = 4$. If so, the sequencer is set to 4, sending it into time step T_4. If not, the machine proceeds to step T_2. In step T_2, we check to see if the addition is required, perform it if so, and then proceed to step T_3. In step T_3 we perform the shifts, decrement I, and proceed back to the start of the loop, at time step T_1. In step T_4, when we exit the loop, the machine does nothing, and halts, allowing the result to be read from the P register.

At this point, we could take the RTL description of the multiplier, and design the data-path and control unit, finishing the design. This process, however, is fairly mechanical, and is left as an exercise for the reader. Before we move on, though, it is worth talking a little bit about how the multiplier is used. From the behavioral description, it is clear that the multiplier is a sequential circuit. This is a little strange. Most computational units of the ALSU are combinational, and can compute their result in one clock cycle. This multiplier, on the other hand, would take 13 cycles to complete a 4-bit calculation: one cycle to initialize the registers, and four iterations of the *while* loop, each iteration involving three clock cycles. In general, if we have a k-bit multiplier, based on this design, the number of clock cycles required to complete the calculation would be $1 + 3k$.

You can see that computer multiplication is significantly slower than computer addition. And, this poses a problem. When performing an addition, the ALSU produces the result in one clock cycle. This is, in fact, how the ALSU is designed to function. But, when performing a multiplication, the ALSU will take several clock cycles. What this means is that when the processor CU requests a multiplication, it must *stall* the machine cycle, for multiple clock cycles, while the processor waits for the result. This stall involves performing a series of phases, that do nothing more than increment the sequencer.

8.2.2.4 Unsigned Division

We start our discussion of computer division with an example. Suppose that we are calculating the value of $1001111 \div 111$. In decimal we are performing the calculation $79 \div 7$, which yields a quotient of 11, and a remainder of 2. (Notice that we get two results from an integer division: the quotient, and the remainder.) To begin with, we show how this calculation would be performed, using long-division, in binary.

$$
\begin{array}{r}
01011 \\
111\overline{)1001111} \\
-111 \\
\hline
010111 \\
-111 \\
\hline
1001 \\
-111 \\
\hline
010
\end{array}
\tag{8.10}
$$

In decimal, long division involves subtracting multiples of the divisor from the dividend. The multiplied divisor is shifted towards the right, as the remaining dividend becomes smaller and smaller. Each multiplier, used to multiply the divisor, is recorded at the top, and becomes one of the digits of the quotient. At the end of the procedure, the remaining dividend is the remainder of the problem.

Calculation 8.10 is this long-division procedure, done in binary. The procedure is simpler in binary, since when we multiply the divisor, we always multiply by one or zero. The result is that our subtractions simply subtract the divisor from the remaining dividend. Just as in decimal, we collect multipliers at the top, to form the quotient. The multiplier bits at the top will be 0, if the shifted divisor is larger than the remainder, or 1, if the shifted divisor is at most as large as the remainder, and the subtraction is performed.

Let us now discuss how this would be implemented in circuitry. We will need three registers: the divisor register, D, the dividend/remainder register, R, and the quotient register, Q. The R register starts out with the dividend

TABLE 8.2   Trace of division,

D	Q	R	Action
01110000	0000	01001111	←
	0000	10011110	-
	0001	00101110	←
	0010	01011100	←
	0100	10111000	-
	0101	01001000	←
	1010	10010000	-
	1011	00100000	✓

as its value, but as mentioned, at the end of the procedure, it will contain the remainder. We will be performing 4-bit division, meaning that an 8-bit dividend will be divided by a 4-bit divisor, producing a 4-bit quotient and a 4-bit remainder. You can think of this as the opposite of multiplication, in which two narrow numbers are combined into a wider number. In division, a wide number is split into two narrower numbers.

If we look at the sizes of our registers, we see that the register R must be eight bits. This is because, initially, it contains the dividend, which is an 8-bit value. The divisor register, D, is being subtracted from the R register, and so it must also be eight bits. The quotient register, Q, needs only to be as large as the quotient we expect, and so it is a 4-bit register. In general, for a k-bit division, both the D and R registers would be size $2k$, and the D register would be size k.

As with multiplication we make a modification with the shifting direction. In normal long-division, the remainder is held fixed, and the divisor is shifted right. We reverse this, and hold the divisor fixed and shift the remainder left. Notice that the first subtractions, in long division, occur with the divisor shifted as far as possible to the left. As a consequence we will start the procedure with the 4-bit divisor shifted to the left of the 8-bit D register, much in the same way as we did in multiplication, for the multiplicand.

Table 8.2 shows the execution of Calculation 8.10, using the registers D, Q, and R. The D register starts with the divisor shifted up to the high-order half of the register. In multiplication, the operations performed in an iteration of the loop were the addition, and then the shift, in that order. In division, the order is reversed. In each iteration, we first shift, and then subtract, if possible. So, in the first two lines of the table, we see the operations performed in the first iteration of the loop. In the first line, both the Q and R registers are shifted left. Then in the second row, the divisor is subtracted from the remainder. The subtraction is performed any time that $R \geq D$. When a subtraction is performed, the low-order bit of the Q register is also changed to a 1. This is what is shown in the second row.

The third row of Table 8.2 represents the second iteration of the loop. The two registers are shifted, and after the shift, $R < D$, so that the subtraction is not performed. The fourth and fifth rows represent the third iteration of the loop. The shift is performed, and a subtraction is possible. The sixth and seventh rows show the fourth iteration, and the last row shows the result. When the procedure terminates, after four iterations, the remainder is found in the high-order half of the R register.

The algorithm used in Table 8.2 is shown below. It starts by shifting the divisor to the high-order half of the D register, initializing the Q register to zero, as well as the loop-counter I. In the loop, we first perform 1-bit left shifts of the Q and R registers, and then perform the subtraction, if $R \geq D$.[1] We can check whether $R \geq D$ by performing the subtraction $R - D$, and then checking the carry-out of the result; if the carry-out bit is 0, indicating an overflow, then $R < D$. At the end of the loop we increment the loop-counter, I. When finished with the loop, we move the remainder from the left of the R register, down to the right.

$$D_{7-4} = D_{3-0}$$
$$D_{3-0} = 0$$
$$Q = 0$$
$$I = 0$$
$$\texttt{while } I \neq 4 \texttt{ do}$$
$$Q = Q << 1$$
$$R = R << 1 \qquad\qquad (8.11)$$
$$\texttt{if } R - D \geq 0 \texttt{ then}$$
$$R = R - D$$
$$Q = Q + 1$$
$$I = I + 1$$
$$R_{3-0} = R_{7-4}$$
$$R_{7-4} = 0$$

---

[1] The divider we have built is typically called a *non-restoring divider*. There are what are called *restoring dividers*. The difference, in the two types, has to do with how the subtraction is done. The non-restoring divider first determines if the subtraction is acceptable, and if so, only then does it perform the subtraction. In a restoring divider, the subtraction is always done in each loop iteration. Only then does the divider check if the subtraction was acceptable, or whether there was an overflow. If there was an overflow, the subtraction is "undone," by adding the divisor back to the remainder.

Algorithm 8.11 can be rewritten as an RTL description, much in the same way as we did for the multiplier described in Algorithm 8.8.

$$Def: I4X \equiv I = 4, T_k \equiv C = k, RDX \equiv R - D \geq 0$$
$$T_0 : D_{7-4} \leftarrow D_{3-0}, D_{3-0} \leftarrow 0, Q \leftarrow 0, I \leftarrow 0, C \leftarrow C + 1$$
$$T_1 \cdot I4X : C \leftarrow 4$$
$$T_1 \cdot \overline{I4X} : C \leftarrow 2$$
$$T_2 : Q \leftarrow shl\ Q, R \leftarrow shlR, C \leftarrow C + 1 \qquad (8.12)$$
$$T_3 \cdot RDX : R \leftarrow R - D, Q \leftarrow Q + 1$$
$$T_3 : I \leftarrow I + 1, C \leftarrow 1$$
$$T_4 : R_{3-0} \leftarrow R_{7-4}, R_{7-4} \leftarrow 0, C \leftarrow 5$$
$$T_5 :$$

Time step T_0 does the initialization, time step T_1 does the loop test, T_2 does the shifting, T_3 does the subtraction if necessary, and increments the loop counter. To decide whether or not to subtract, we use the signal RDX, which is the result of the comparison $R \geq D$. Step T_4 shifts the remainder right, and T_5 stalls the divider, with the final result.

As with the multiplier, we leave the development of a structural description to you, as an exercise. It should also be fairly obvious how to extend our design from 4-bit division to k-bit division, and so we are done with division. Before we finish up with the divider, however, we would like to return to a statement we made at the beginning of this section. The statement was that the same circuit that was used for the multiplier could also be used as the divider.

In Section 8.2.2.2, we made the claim that computers do not perform subtraction, only addition. We went on to explain that what we mean by this is that the same circuitry performs both addition and subtraction. The same statement can be made about multiplication and division. That is to say that the circuitry that performs multiplication, can also be reused to perform division.

We typically refer to the multiplier circuit we designed as the *add-shift* multiplier. In this same way, we refer to the divider as the *shift-subtract* divider. The names of these devices emphasize that they are very similar. The multiplier performs addition, and sifts an operand right. The divider performs subtraction, and shifts the operand left. If we were to construct a device that was "reversible," we could use it to do both multiplication and division. When we say that the device is reversible, we mean that we can reverse the direction of the shift, invert the addition to subtraction, and invert the order of the shift and addition operation. Such a reversible device is, in fact, possible to build. The advantage of using a single device to perform both multiplication and division, is that we then decrease the size of the processor, with respect to the situation where we used two distinct devices.

8.2.3 Signed Arithmetic

We now turn to the task of building computational units to perform signed arithmetic. We will consider the same set of arithmetic operations, as we did for unsigned integers: addition, subtraction, multiplication, and division.

8.2.3.1 Signed Addition and Subtraction

When discussing unsigned subtraction in Section 8.2.2.2, we already performed addition on some negative numbers. And, you will have observed that we simply added them, in exactly the same way as we would add positive numbers. What we learn from this is that the same circuit that performs unsigned addition, the adder, also performs signed addition. This is excellent news; we do not have to add circuitry to the processor to accommodate signed addition.

Remember that in unsigned operation, there is also no extra circuitry needed for subtraction. To perform subtraction, we simply add the two's compliment of the subtrahend to the minuend. This same method of subtraction is used with signed numbers. The result of this discussion is that we have now realized that to perform all addition and subtraction, only one computational unit, the adder, is needed.

One difference between signed and unsigned arithmetic, however, is the handling of the overflow flag. Let us examine this subject with a few examples. We start with the 8-bit example 01110110 + 11011010.

$$
\begin{array}{r}
{\scriptstyle 111111100} \\
01110110 \\
+11011010 \\
\hline
01010000
\end{array}
\tag{8.13}
$$

Calculation 8.13 represents the calculation $118 - 38 = 80$. If you convert the result of Calculation 8.13 into decimal, the result is, in fact, 80, and so this calculation does not result in overflow, and this is what would have been predicted by our overflow rule, Equation 8.6, for subtraction of unsigned numbers.

As another example, consider the signed calculation 01011111 + 00110011. In decimal, this would be $95 + 51 = 146$.

$$
\begin{array}{r}
{\scriptstyle 011111110} \\
01011111 \\
+00110011 \\
\hline
10010010
\end{array}
\tag{8.14}
$$

You should immediately notice that there is something wrong with this calculation, since two positive numbers are being added, and the result is negative

(−110). This is overflow, and it is the result of a sum that is too large to be stored in 8-bit signed format. (Remember that the largest signed integer we can store in an 8-bit format is $2^7 - 1 = 127$.) But, something interesting has happened with the carry-out. The carry-out is 0, which, for unsigned addition, would indicate that there is no overflow, as specified in Equation 8.3. Clearly we need a different definition of overflow for signed addition.

Our definition of overflow for signed addition is based on our observation that in Calculation 8.14 we had two operands of the same sign, and the sum was of the opposite sign. In fact, this is one legitimate definition of overflow, for signed integer addition. Notice that when the signs of the addends are different, as in Calculation 8.13, we did not get overflow, and actually, overflow is impossible in this case. The reason is that the sum will be smaller in magnitude than the addends, and if the addends are small enough to be represented in eight bits, so also will be the sum. The only way to produce a sum large enough to exceed the word size is if the two addends are of the same sign.

Although we have a working definition of overflow in signed addition, it is a bit difficult to implement in circuitry, or at least there is a simpler, but equivalent, way of checking for overflow with circuitry. For signed addition we define overflow as follows.

$$V = C_{out} \oplus C_{in} \tag{8.15}$$

Here, the carry-in and carry-out we are referring to are the carry-in and carry-out of the sign bit. Remember that the XOR function, with two operands, checks if the two operands are not equal. In other words, an overflow occurs when the carry-in and carry-out of the sign bit differ. If we apply this rule to Calculation 8.13, we see that $C_{in} = 1$ and $C_{out} = 1$, and so $C_{in} = C_{out}$, indicating that there is no overflow. In Calculation 8.14, $C_{in} = 1 \neq 0 = C_{out}$, and so there was overflow.

One more example demonstrates our overflow rule with negative addends.

$$
\begin{array}{r}
{\scriptstyle 111110000} \\
11011000 \\
+11111101 \\
\hline
11010101
\end{array}
\tag{8.16}
$$

In Calculation 8.16, we do the calculation $-40 - 3 = -43$. Notice that we have a carry-out of 1, but that this does not mean overflow, for signed addition. Instead, we observe that both carry-in and carry-out are 1, indicating no overflow.

8.2.3.2 Signed Multiplication and Division

We now turn to the multiplication and division of signed integers. We already have a multiplier, to multiply unsigned integers, and a divider circuit, to divide unsigned integers. We could use these circuits to manipulate signed numbers also. For example, if we wished to multiply the two signed integers, 11110110 and 00000110, we could do the following.

1. Calculate their magnitudes: 00001010, and 00000110.

2. Use the unsigned multiplier to multiply the magnitudes: $0001010 \times 0000110 = 00111100$.

3. Calculate the product sign, by taking the XOR of the two operand sign bits[2]: $1 \oplus 0 = 1$.

4. If the product sign is negative, take the two' s compliment of the product magnitude, to produce the actual product: 11000100.

So, here we see that the multiplier/divider circuit we have developed can be used to operate on signed numbers, as well as unsigned numbers, simply by using it to operate on the magnitudes of the operands. Another option is to try and develop a circuit that could operate, directly, on signed numbers. Such a circuit, called a *Booth's* multiplier,[3] exists for multiplication, but regrettably, not for division. This means that, if we wish to use a Booth's multiplier, our ALSU must contain a multiplier, and a separate circuit for division, whereas, if we use the standard add-shift multiplier, the same circuit can be used for both multiplication and division.

Returning to the add-shift multiplier, we now know how to calculate the sign of a product: by taking the XOR of the operand sign bits. The next question is how to determine the sign of the results of a division. Notice that, in division, we end up with two results: the quotient and the remainder. We must discuss how to determine the sign of both results.

If we use the shift-subtract divider to divide the magnitudes, we are faced with calculating the sign of the quotient. You should be able to verify that the sign of the quotient follows the same rule as the sign of the product of an add-shift multiplication. That is to say, the sign bit of the quotient is just the XOR of the sign bits of the dividend and the divisor.

---

[2]The sign of the product would be 1 (negative), if the signs of the multiplicand and the multiplier are not equal, and would be a 0, if the signs were equal. In other words, the sign bit of the product is the XOR of the sign bits of the operands.

[3]A Booth's multiplier is based on the *Booth's expansion* of a binary number. In normal binary expansion we see a binary number as the sum of a sequence of powers of two. For example the binary number 00011111 expands to the sequence $2^4 + 2^3 + 2^2 + 2^1 + 2^0$. Booth observed that a string of contiguous ones can be represented by a subtraction and an addition. So, the Booth's expansion of the number 00011111 would be $2^5 - 2^0$. A multiplier that uses Booth's expansion would both add, and subtract shifted copies of the multiplicand. It is from the inclusion of subtractions that Booth's algorithm procures its ability to handle operands and products that are both positive and negative.

In terms of the remainder, there is a slight issue that must be addressed, when we start dealing with signs. The issue can easily be demonstrated with an example. Suppose that we are doing the calculation $-56 \div 9$. If we use true integer division to do this problem, we get $\lfloor -56/9 \rfloor = -7$, and we could then write the dividend as the quotient multiplied by the divisor, and added to the remainder, as follows: $-56 = 9 \cdot (-7) + 7$. If, however, we do what we have proposed to do, and use the shift-subtract divider to divide the magnitudes, we would find that $\lfloor 56/9 \rfloor = 6$, and so our signed quotient would be -6, resulting in an expansion of $-56 = 9 \cdot (-6) - 2$.

We have discovered that the above calculation has two possible answers, depending on how we implement computer division. Which of the two implementations is "more correct" is not clear-cut. To determine which is better, it is probably a good idea to resolve the issue in a way that a human user would find intuitive. We call the remainder the "remainder," for a reason. People tend to view the remainder as the part of the dividend that "remains" when we subtract out the quotient, multiplied by the divisor. Because of this, we usually expect the remainder to have the same sign as the dividend. And, this is the rule that is, typically, used in computer arithmetic. This corresponds to the definition of integer division as that resulting from dividing the magnitudes of the operands, using the shift-subtract divider.

8.2.4 Floating-Point Data

We use floating-point data to represent real numbers. There is, of course, a problem with trying to represent arbitrary real numbers, in particular irrational numbers, on a computer. Suppose that you decide to represent real numbers using 32 bits. Now, you examine the number π. As you know, π is irrational. If you look at its decimal expansion, it is infinite in length. The same is true for its binary expansion. What is more, the expansion is non-repeating. This is a problem that, to solve, will require a representation with an unbounded number of bits. Certainly, 32 bits will not suffice.

As a consequence of the above argument, when we say that floating-point is used to represent real numbers, we must admit that we can only represent some real numbers exactly. In particular, we can only represent rational numbers, and even then, only some rational numbers. To represent an irrational number, like π, what we do in floating-point notation is to store an approximation, meaning a rational number with a finite-length expansion that is fairly close to π.

The next question is how do we store a rational number, with a finite expansion. For the purposes of this discussion we will work in decimal. Consider the rational number -43.8125. We learn early on in our math education that any such number can be written in *scientific notation*, a standard way of writing any rational number. For this example, we would write the number as -4.38125×10^1, in scientific notation. The number is written as a *mantissa* multiplied by a base (base 10 in decimal), raised to a power, called the *exponent*. To form the mantissa, we move, or *float*, the decimal point, in order to *normalize* the mantissa. This is why we call this type of representation floating-point.

A mantissa is considered normalized if there is only one digit preceding the decimal point, and that digit is in the range 1 through 9. (Zero is not allowed as the lead digit.) Notice that there are rational numbers that cannot be normalized. In particular, you might, with futility, try to normalize the number 0. Clearly, you cannot normalize zero, because the leading digit can only be 0. The best we can do for zero is to come up with some kind of convention on how it would be written in scientific notation, that is an exception to the normalization rule. The convention might be to write zero as 0.0×10^0.

There are, at least, three pieces of significant information contained in a scientific notation representation of a rational number. We have already mentioned two of them: the mantissa and the exponent. The third piece of information that is significant is the sign of the number. There are, in addition, two other pieces of information contained in the notation, that we do not consider significant: the sign of the exponent and the base. The base is unimportant, because in decimal scientific notation we always use ten as the base. Unless we are changing bases, this is understood. The sign of the exponent can be considered as part of the exponent, and although significant, we would not consider it as a separate piece of information.

We use scientific notation to represent rational numbers on a computer. We refer to scientific notation as floating-point notation, for reasons already touched on. Of course, on a computer, we will be using binary scientific notation. Binary floating-point notation is similar to decimal floating-point notation. As an example the binary number -101011.1101 can be written in floating-point as -1.010111101×2^5. In binary the base is two. You will notice that the number has been normalized. In binary, a normalized mantissa must begin with the digit 1, before the binary point, since the leading digit is not allowed to be 0.

Given that we are using binary floating-point to represent rational numbers, we now must come up with a scheme for storing our floating-point representation on a computer. The first decision we make is to decide on the width of the *floating-point word*. This is the storage unit that will be used to store floating-point numbers. The floating-point word may, or may not be the same width as the integer word for the processor, and we will talk more about this a little later, in Section 8.2.4.2. Once the size of the floating-point word has been determined, we then split the word into three fields. Each field is used to store one of the three significant pieces of information. So, we will have a sign field, a mantissa field, and an exponent field.

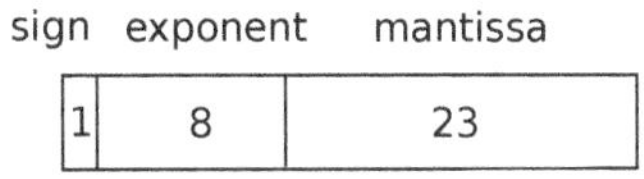

sign exponent   mantissa

1	8	23

FIGURE 8.3 Format of a 32-bit floating-point word.

Figure 8.3 shows a possible format of a 32-bit floating-point word. The word is split into the three fields. The high-order bit of the word, following the same convention we use for the signed integer word, is the sign bit, with a 0 bit representing a non-negative number, and a 1 bit representing a negative number. The next field is the exponent field, which is allocated eight bits. The last field is the mantissa field, which is allocated the remaining 23 bits.

We can store mantissas with up to 23 bits of precision. But, using the fact, already discussed, that the leading digit in a normalized mantissa is always 1 in binary, we can actually store 24 bits of precision. We do this by making the leading digit implicit. For our example, -1.010111101×2^5, we would drop the digit before the binary point, and the binary point, and store only the digits after the point, padded out to 23 bits: 0101 1110 1000 0000 0000 000. This makes our representation for floating-point numbers a little harder to use; when we pack a number into a floating-point word, we must truncate off the leading one, and when we unpack a floating-point number, for use in arithmetic, we must restore the leading digit.

Although what we do to the mantissa is a little unusual, what we do to the exponent field is really strange. Remember that we said that the sign of the exponent is incorporated into the exponent. To do this we might use two's compliment notation to store signed exponents. But, that is not what is done. Instead, we store the exponent in what is called *bias* notation. In bias notation, the sign bit of the exponent functions in reverse. That is to say that a sign bit of 0 indicates a non-positive number, and a sign bit of 1 indicates a positive number. If we count in 8-bit bias notation, also called 127-bias notation, starting at the lowest possible number, up to the highest possible number, we get the values shown in Table 8.3.

TABLE 8.3   127-bias exponent values.

Decimal Number	127-Bias
−127	00000000
−126	00000001
−125	00000010
−124	00000011
⋮	⋮
0	01111111
1	10000000
2	10000001
3	10000010
⋮	⋮
128	11111111

To convert an exponent into its 127-bias notation looks complicated, but actually, it is quite simple; you simply add 127 to the exponent. For example, if the exponent is 0 (00000000 in 8 bits), you add 127 (01111111 in 8-bit binary), and get 01111111. If the exponent is −5, you compute −5 + 127 = 122, and write 122 in 8-bit binary as 01111010. Converting from 127-bias notation to the exponent is just as easy; you simply take the 127-bias number, convert it to decimal, and subtract 127 from that. For example, if your 127-bias exponent is 10000111, you convert that to 135, subtract 127 from this, and arrive at 8, as the effective exponent.

There is even an easier way to convert to/from 127-bias from/to a decimal exponent. It involves two copy transformations. One transformation is used for positive numbers, and one transformation is used for non-positive numbers. We will illustrate the technique with two examples.

We start with an example for a positive exponent. Suppose that your exponent is 12, in decimal. To write twelve in 127-bias notation you start by writing down the sign bit for positive numbers, 1, and follow this with a 7-bit eleven. The result is 1 0001011. In other words, the exponent is copied into seven bits, after subtracting a one from it.

For non-positive numbers, we take the exponent −12 as a example. In the copy transformation for non-positive exponents, you again start by writing down the sign bit, (0 for non-positive) and then follow this with the 7-bit one's compliment of the magnitude of the exponent. The result for −12 is 0 1110011, where 1110011 is the one's compliment of the 7-bit binary representation of 12, 0001100.

TABLE 8.4   Successive multiplication conversion.

Calculation	Integer Part	Fractional Part
.8125 × 2	1	.625
.625 × 2	1	.25
.25 × 2	0	.5
.5 × 2	1	.0

Returning to our original example, -1.010111101×2^5, based on the discussions we have just presented, this number would be stored in the floating-point word as follows: a sign bit of 1, an exponent of 10000100, and a mantissa of 0101 1110 1000 0000 0000 000. The result is 1,10000100,0101 1110 1000 0000 0000 000.

8.2.4.1   Converting between Floating-Point and Decimal

With the knowledge we now have on floating-point format, we should feel capable of freely converting between floating-point format and decimal expansion. We will present a couple of examples to convince you of this. First, we present an example converting from decimal to floating-point. Consider the number –43.8125. We wish to write this number into a floating-point word. To do this, we follow a three-step process.

1. Convert the decimal number to binary.

2. Write the binary number in scientific notation.

3. Pack the scientific notation number into the floating-point word.

For our example, we start by converting the decimal number –43.8125 into binary. We should be able to easily convert the integer part of the number, 43, into binary: 101011. But then, we must also convert the fraction into binary. To do this we use the method of *successive multiplication*, which is a technique similar to successive division, used to convert an integer into binary, and covered in Section 2.1. In successive multiplication, we multiply the fraction by two, on each pass. This gives us two results: a new fraction, and an integer part of the result. The integer part is either a 0 or a 1. We use these bits, from all of the passes, to form the binary representation of the fraction.

Table 8.4 shows the calculation of the binary fraction for our example. We start by multiplying the fraction by two, producing the result 1.625. The integer part, 1, is written in the second column, and the fractional part, .625, is written in the third column. We then take this new fraction and multiply it by two, in the second row, producing a new fraction, .25, and new integer, 1. We continue this process on each row. The process terminates when we get a fraction .0. (The process may not terminate, in some cases, indicating that the binary expansion of the fraction is infinite in length.) To form our binary fraction, we observe that each time we do a multiplication by two, one binary digit hops over the binary point, and shows up as the integer part. The first digit that hops over the point is the digit closest to the binary point, in the binary expansion. So, to form the binary expansion of the fraction we would read the integer bits from Table 8.4, top to bottom, and write them left to right. This gives us the binary fraction .1101. The binary form of -43.8125 is then -101011.1101.

To convert into floating-point, our second step is to write our binary number in scientific notation. For our example, this would be -1.010111101×2^5. In the third step, we would then pack this into a floating-point word. For this example, which is a negative number, the sign bit would be 1; the exponent, 5, in 127-bias notation would be 10000100; and the mantissa would be 0101 1110 1000 0000 0000 000, after truncating the leading bit, and padding it out to 23 bits. The final result is 1,10000100,0101 1110 1000 0000 0000 000.

Now, let us reverse the process. That is, we will start with the floating-point representation 0,01111110,1010 0000 0000 0000 0000 000, and convert it into a decimal representation. The procedure we follow is the reverse of the procedure we just used to convert from decimal to floating-point. Examining the floating-point word, the sign bit, 0, indicates a non-negative number. The 127-bias exponent is 126 in decimal, and subtracting 127, this gives an effective exponent of -1. The mantissa, after adding back the truncated 1 and the binary point, is 1.101, resulting in the scientific notational representation of $+1.101 \times 2^{-1}$.

Once we have the scientific notational representation, we have two alternative ways in which we can proceed. We often refer to these as the *convert and multiply* (CM) method, or the *multiply and convert* method (MC). We will illustrate both methods, starting with CM.

In the CM method, we first convert all parts of the scientific notation into decimal fractions, and then perform the multiplication. For our example, it is easy to convert the power, 2^{-1}, to a fraction: $\frac{1}{2}$. To convert the mantissa into a decimal fraction, we use a trick we often use in decimal. For example, in decimal, the number 0.349 can be written as $\frac{349}{1,000}$. You do this by reading the fractional number as an integer, by ignoring the decimal point. This gives us the numerator 349. The denominator is the power of ten corresponding to the position of the low-order digit of the fraction. For our example, $0.349 = 3 \times 10^{-1} + 4 \times 10^{-2} + 9 \times 10^{-3}$, and so the low-order digit, 9, corresponds to the power $10^{-3} = \frac{1}{1000}$, which becomes the denominator. Using the same trick in binary, the number $1.101 = \frac{1101}{2^3} = \frac{13}{8}$. Our full binary number is then

$\frac{13}{8} \times \frac{1}{2}$, after converting everything to decimal. After conversion to decimal, in CM, we would perform the multiplication, yielding $\frac{13}{16}$, and then finish by doing the division, yielding 0.8125. Combining the result with the sign, we arrive at our final result: +0.8125.

In the MC method, we first multiply the mantissa by the power. In our example this calculation is $1.101 \times 2^{-1} = 0.1101$. Then we convert the resulting binary number into decimal, yielding $0.1101 = \frac{1101}{2^4} = \frac{13}{16}$. From here it is a simple matter of doing the division, and adding in the sign, to get the same result as in the CM method.

We should now have a fairly good grasp of the floating-point notation, and how to interpret it in decimal. There are a few issues with this representation that it would be remiss not to discuss.

8.2.4.2 Standardization

Historically, different computer manufacturers were using different floating-point formats for the floating-point word. They differed in word width, field width, field position, and field encoding. This situation created several problems, and people found that the format differences would result in floating-point code that would work on one type of machine, but not on anther model. This lack of code portability was often the result of a field, like the exponent field, being wider on one machine, than on another. With a wider field on one machine, the code would work, but on another machine, with a narrower field, the calculations would result in overflow.

To solve the problems of portability, industry stakeholders developed a standard format for floating-point representation. This standard, known as IEEE 754, was so successful, that today it is so widely implemented, that it is taken for granted. We have been using the IEEE 754 standard in our discussion. It uses a 32-bit floating-point word, with an 8-bit exponent. Figure 8.3 shows the single precision format for the IEEE 754 standard.

With an 8-bit exponent, we have already pointed out that you can represent exponents with magnitudes up to about 127. With the 23-bit mantissa, we can also approximate numbers with a great deal of precision. You might think that this 32-bit representation would be more than adequate. But, there are many scientists doing calculations with floating-point numbers, and these calculations can be very sensitive, and need to be very precise. So as it turns out, the 32-bit format is not always sufficient. This fact was not lost on the developers of the IEEE standard. To address this issue, they actually developed two floating-point formats, that are supported on most machines: a 32-bit format, called *single precision* format, and a 64-bit format, called *double precision* format. In the language *C*, you have the two real types, `float` and `double`, that support these two formats.

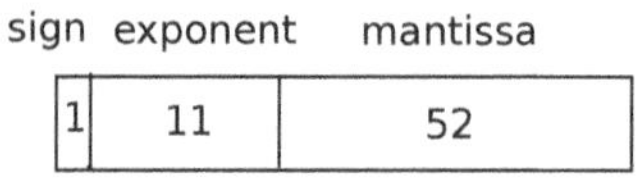

sign exponent   mantissa

1	11	52

FIGURE 8.4 Format of a 64-bit floating-point word.

Figure 8.4 shows the double precision format for the IEEE 754 standard. Numbers are packed into this floating-point word in the same way as in the single precision word. That is to say, you eliminate the leading digit of the mantissa, and the exponent is stored in bias notation. One difference is that the double precision format uses 1023-bias notation, rather than 127-bias notation.[4] The double precision format provides a 52-bit mantissa field, allowing almost twice the precision, and an 11-bit exponent, allowing exponents of up to 1023.

In the IEEE 754, usually mantissas would be normalized. We have, however, already mentioned that the number zero cannot be normalized. The best we can do is standardize its representation. The developers of the IEEE standard decided to represent it with the same bit configuration used to represent the integer zero: 0,00000000,0000 0000 0000 0000 0000 000.

There are a few other non-normalized numbers in the IEEE standard. These are numbers indicating special conditions in arithmetic calculations. They include negative infinity, positive infinity, and a number with the rather ironic name *not-a-number* (NaN), often used to indicate an undefined result.

8.2.4.3 Field Order

One non-intuitive part of the IEEE floating-point standard is the field order. When we write a real number in scientific notation, we write the mantissa first, followed by the exponent, as in the example -5.678×10^9. And yet, the developers of the IEEE 754 standard chose to arrange the fields with the exponent first, and then followed by the mantissa. So, the question is why this was done.

---

[4]The bias is determined by the field width. The bias is always the bit configuration representing zero. This value is always a leading 0, followed by a sequence of ones. For 127-bias, with an 8-bit field, this is 01111111 (127 in decimal). For an exponent of eleven bits, the bias would be 01111111111, which is 1023 in decimal.

The question is answered, when we start comparing floating-point numbers. For example, suppose we wish to determine if the floating-point number $a = $ 0,10000011,0010 0000 0000 0000 0000 000 is larger than the floating-point number $b = $ 0,01111000,0011 0000 0000 0000 0000 000. It would be nice if we could use the same comparator circuit to make this determination, as we would use for integer data. If you look at the two bit strings as integers, and compare them, you find that $a > b$, which is actually the same determination you would make if you compared a, and b as real numbers. However, if you reverse the order of the exponents and mantissas, so that $a = $ 0,0010 0000 0000 0000 0000 000,10000011, and $b = $ 0,0011 0000 0000 0000 0000 000,01111000, comparing the two values as integers, you find that $b > a$. The lesson here is that if you wish to compare floating-point numbers, as integers, you must ensure that the most significant bits of the floating-point number fall in the most significant positions of the integer word. With floating-point numbers, the exponent is much more significant than the mantissa at determining the magnitude of the number, and so in the IEEE 754 standard, the exponent is placed before the mantissa.

We would like to mention, also, that the reason for the order of the fields in the floating-point format is also the reason we use bias notation for the exponent. If we use integer comparison on floating-point numbers, we must ensure that the bit configuration of a positive exponent, interpreted as an unsigned integer, appears larger than that of a negative exponent. This is not the case with two's compliment notation, in which a negative exponent starts with a 1 in the sign bit, and a positive exponent starts with a 0. It is, however, true of bias notation, where a negative exponent starts with a 0, and a positive exponent starts with a 1.

8.2.4.4 Arithmetic Approximation

As previously mentioned, we really cannot represent arbitrary real numbers using floating-point representation. The best we can do is approximate them with rational numbers with finite-length expansions. This statement might start us wondering how good an approximation we can achieve.

To answer the question of how good our approximation is, we look at three properties that are a direct result of the way in which we are representing floating-point numbers.

- *Precision.* The number of digits in the mantissa. So, for the IEEE single precision format, we would say that we have a precision of 24 bits. (This works out to be about seven decimal digits of precision.)

- *Range.* This is the interval of numbers that we can represent, from the largest to the smallest. For the single precision format, this is the interval

[1,11111110,1111 1111 1111 1111 1111 111 .. 0,11111110,1111 1111 1111
1111 1111 111] (the smallest non-infinite negative number, to the largest
non-infinite positive number).

- *Gap.* The largest distance between consecutive, representable rational
  numbers.

The precision and range are concepts that are probably familiar to you
in other contexts. The gap, however, requires a little explanation. What
is necessary to understand is that not all real numbers in the range can
be represented in floating-point. For example, take the floating-point number 0,10000011,0101 0000 1111 0101 0000 110, which represents the number
$+1.0101, 0000, 1111, 0101, 0000, 110 \times 2^4$ (approximately 21.059826). The next
largest real number that it is possible to represent is 0,10000011,0101 0000
1111 0101 0000 111, or $+1.0101, 0000, 1111, 0101, 000, 111 \times 2^4$ (approximately
21.059828), derived from the first number by flipping the rightmost bit of the
mantissa. The point is that there are a lot of real numbers (actually an infinite
number) between these two numbers, and none of them have a floating-point
representation. So, the gap refers to this discontinuity, and in particular to
the largest such "hole" between consecutive floating-point representations.

The three properties—precision, range, and gap—determine how well we
can approximate an arbitrary real number. Obviously, we would like a large
precision and a small gap. With a small gap, if the real number that we wish to
approximate falls within the range, we can then come up with a fairly precise
approximation to the number. If the range is large, we have a good chance of
our number being within the range. But, no matter how large the range, there
are numbers outside of the range, and we are bound to do some arithmetic
calculation, on some occasion, that produces a result that is one of these
extreme numbers. If this happens, the ALU needs to signal the processor, so
that this problem can be handled.

We have dealt with this issue, when we talked about integers and *overflow*.
This situation, which we are discussing, is floating-point overflow. Just as for
integer overflow, we now ask ourselves how to detect floating-point overflow.
The answer has to do with examining the most significant part of the floating-
point number: the exponent. To detect overflow after a calculation, what we
do is attempt to pack the exponent into bias notation. If the exponent is
too large to pack (in 127-bias notation, it is larger than 127), then we have
overflow.

In addition to overflow, floating-point calculation can also result in *underflow*. Overflow occurs when the magnitude of a result is too large. Underflow
occurs when the magnitude of the result is too small, or stated another way,

the result is too close to zero. This happens when the resulting exponent is below –127, for single precision.

8.2.4.5 Rounding

As we know, many floating-point numbers are approximations. And, you are probably not surprised that the way we approximate an arbitrary real number is with the closest floating-point number, either in the positive direction, or in the negative direction. The choice of which of these two options to use is encompassed in the topic of *rounding*.

To introduce rounding, we will present some examples. Rather than work with IEEE 754 single precision, which, with 24 bits of precision, is a bit clumsy for our purposes, we will work with only five bits of precision. So, we might have a floating-point number like $+1.0111 \times 2^6$, written in scientific notation. We do arithmetic calculations with these floating-point numbers. Often, we get, as results, numbers that require more precision, like $+1.011011 \times 2^6$, which requires seven places of precision.

When we perform arithmetic, we cannot keep infinite precision in our answers. We should readily realize that we will have to discard bits in our answers. As a matter of fact, it is usual to discard all but two extra bits. That is to say, if we have a floating-point implementation with five places of accuracy, we would maintain the five bits, and two extra bits, in our intermediate answers, for a total of seven bits of precision. The number $+1.0110, 10 \times 2^6$ might easily be the result of a computation on such a system. Here we separate the extra two bits from the rest of the mantissa with a comma. The two extra bits are usually referred to as the *round bit*, which is the high-order bit, 1, and the *guard bit*, which is the lower-order bit, 0.

We use the round bit to round the result up to, or down to the nearest floating-point number. The guard bit is used in rounding, but that is not its major purpose. We will find that while performing arithmetic, we often shift right, and later shift left. This is a direct result of our requirement for normalization. As we shift right, we lose bits off the low end of the mantissa. Later when we shift left, without the round and guard bits, the lost bits would be replaced by zeros. But the precision of the lost bits may be needed in future calculations. The guard bit gives us a way of saving an extra digit during the right shift, increasing our accuracy.[5]

---

[5] Many floating-point implementations keep a third bit, called the *sticky bit*, that is also used to catch shifted bits.

TABLE 8.5 Rounding method examples.

Method	+1.0110, 10	−1.0110, 10	+1.0110, 01	−1.0110, 01
RN	+1.0111	−1.0111	+1.0110	−1.0110
RZ	+1.0110	−1.0110	+1.0110	−1.0110
RP	+1.0111	−1.0110	+1.0111	−1.0110
RM	+1.0110	−1.0111	+1.0110	−1.0111

The IEEE 754 standard specifies how the round and guard bits are used in rounding. As a matter of fact, the standard gives four methods of rounding. It is up to the processor designer to choose which method to use.[6] The four methods of rounding are the following.

- *Round Nearest* (RN): Round the number to the nearest floating-point number.

- *Round Zero* (RZ): Round the number to the closest floating-point number, towards zero.

- *Round Positive* (RP): Round the number to the closest floating-point number, in the direction of positive infinity.

- *Round Minus* (RM): Round the number to the closest floating-point number in the direction of negative infinity.

Table 8.5 demonstrates how the four rounding methods work on several mantissa examples. Each method performs either a truncate, in which the extra bits are just truncated off of the result, or what we will call a *round up*, meaning that the magnitude of the mantissa is truncated, and then incremented. Starting with RN, in the first row, the first two columns show an example where the number is rounded up. This is because the round bit is a 1. RN is fairly simple, in this respect. If the round bit is a 1, then the number is rounded up, and if the round bit is 0, the number is truncated. Truncation is what happens in the last two columns, where the extra bits are 01.

For RZ, you decrease the magnitude, to make it closer to zero. In RZ, therefore, you always truncate. This is what is happening in the second row of Table 8.5. For RP, you are trying to move the number closer to positive infinity. What this amounts to is that negative numbers are truncated, and positive numbers are rounded up, as shown in the third row. Finally, for RM you move the number closer to negative infinity, by truncating positive numbers and rounding negative numbers up. It goes without saying that if you are lucky

---

[6]The flexibility to choose the rounding method can introduce code portability problems. However, the differences in arithmetic computations are usually so small as to be insignificant.

enough to perform a calculation with a result that is exactly a floating-point number, like a mantissa of $1.0110, 00$, rounding consists of simply truncating the extra bits.

8.2.4.6 Floating-Point Addition

We are now ready to explore floating-point arithmetic. We start with addition, and begin by describing the floating-point adder. Our description of the adder is limited to the algorithm, since by now conversion into RTL, and thence into circuitry, should be a fairly familiar process.

To add two floating-point numbers together, we employ a method derived from the familiar method you have learned for adding numbers in scientific notation. We illustrate this process with an example, and show the results of the steps of the algorithm. The example we will use is $+1.0111 \times 2^3 - 1.1011 \times 2^2$. The steps of the algorithm follow. We fully acknowledge that the presented algorithm would have to be modified to handle non-normalized numbers, but in the interest of simplicity we will ignore their existence. We justify this simplification with the knowledge that the modifications to handle the non-normal numbers are fairly simple.[7]

1. Align binary points. In this step, we ensure that the two addends have the same exponent. It is typical, on a computer, to write both operands with the larger of the two exponents. After the execution of this step, our example has been transformed into the problem

$$+1.0111, 00 \times 2^3 - 0.1101, 10 \times 2^3.$$

Notice how this is being done. The addend with the smallest exponent has its exponent increased to make it equal to the exponent of the other addend. At the same time, the mantissa of the addend is shifted right, by the amount of increase in the exponent. Also notice the inclusion of the two extra bits, separated by a comma, used to catch some of the bits being shifted off. The round and guard bits are kept active, until the end of the algorithm.

2. Add the mantissas. In this step we add the two mantissas, using a signed integer adder. The calculation performed by the adder is shown below.

$$
\begin{array}{r}
001.0111, 00 \\
-000.1101.10 \\
\end{array}
\Rightarrow
\begin{array}{r}
{\scriptstyle 1110\ 1100\ 00} \\
001.0111, 00 \\
+111.0010, 10 \\
\hline
000.1001, 10 \\
\end{array}
\qquad (8.17)
$$

---

[7] Adding zero to any addend, just produce the addend as the result. Adding NaN, negative infinity, or positive infinity to any addend just produces NaN, negative infinity, or positive infinity, respectively.

Calculation 8.17 performs subtraction using two's compliment addition, as the adder would do. We include a high-order bit to accommodate a sign, and an extra bit above the binary point for increased magnitude. The two's compliment result of the adder is converted to sign-magnitude form, so that after this step, the result for our example is

$$+0.1001, 10 \times 2^3.$$

3. Normalize the mantissa. In this step, we shift the mantissa back to normal form. At the same time, the exponent must be adjusted to reflect the shift. It is possible that the mantissa would need to be shifted to the right, for example if the result of an addition had a mantissa of 10.0101,01, or to the left, as in our example. For our example, the result after normalization would be

$$+1.0011, 00 \times 2^2.$$

4. Round the mantissa. In this step, we use one of the rounding methods, discussed previously, to eliminate the extra precision bits. For our example, it does not matter which method we use, since the result, at this point, has zeros in both the round, and guard bits. So, our result would then be transformed into

$$+1.0011 \times 2^2.$$

It is worth mentioning, that after this step, it may be necessary to re-normalize the mantissa. For instance, if your unrounded result is 1.1111,11, and you are performing RP rounding, which requires you to round the mantissa up, after rounding, you would get 10.000, which is not normal, and would require re-normalization.[8]

EZ = EA	1
if EB > EA **then**	2
EZ = EB	3
while EA < EZ **do**	4
EA = EA + 1	5
MA = MA >> 1	6
while EB < EZ **do**	7
EB = EB + 1	8
MB = MB >> 1	9
if SA **then**	10

---

[8]You might worry that you could get trapped in a never-ending cycle; normalize, round, normalize, round, etc. But, it can be shown that this will not happen. The procedure will terminate after at most a normalization, a round, and a single re-normalization.

```
            MA = –MA                          11
    if  SB  then                              12
            MB = –MB                          13
    MZ = MA + MB                              14
    SZ = 0                                    15
    if  MZ < 0  then                          16
            MZ = –MZ                          17
            SZ = 1                            18
    if  MZ[1]  then                           19
            MZ = MZ >> 1                      20
            EZ = EZ + 1                       21
    while  !M[0]  do                          22
            MZ = MZ << 1                      23
            EZ = EZ – 1                       24
    M[ – 5.. – 6] = [0 0]                     25
```

Listing 8.1 Pseudo-code for floating-point adder.

We now take this above, abstract sequence of steps, and convert it into an algorithm. In this algorithm, we assume that the two floating-point operands have already been unpacked into six registers: SA, EA, and MA, which contain the sign, exponent, and mantissa, respectively, for the first, or A operand, and SB, EB, and MB, which contain the sign, exponent, and mantissa, for the second, or B, operand. The result will be stored in the three registers, SZ, EZ, and MZ. The algorithm is set up for our floating-point numbers with five places of precision, but could easily be adapted to the 32-bit precision of the IEEE 754 standard. Our algorithm is presented in Listing 8.1.

In Listing 8.1, we start out by finding the larger of the two exponents in Lines 1 through 3. Lines 4 through 9 align the binary points. There are two *while* loops: one that shifts the A mantissa, if needed, and one that shifts the B mantissa, if needed. Only one *while* loop will be executed, depending on which operand has the smaller exponent.

Lines 10 through 13 replace the mantissas with their two's compliment, if their sign indicates a negative number. Then, in Line 14 the mantissas are added. In Lines 15 through 18, the sign of the result, Z, is calculated, and if the result is negative, the mantissa of Z is replaced by its two's compliment, which is its positive magnitude.

If the result, at this point, is normalized, the bits left of the binary point should be 001, an integer value of one. In Line 19, we check the second bit position above the binary point. (It should be zero.) If it is a one, we shift the mantissa of Z right one bit, in Lines 20 and 21. Note that we number the bits in the register MZ, by their position, relative to the binary point. Positions going left from the binary point are numbered 0 on up, proceeding to the left, as is the usual way of numbering bits for integers. Bits proceeding right from the point are numbered starting with –1, on down. So, the second bit above the binary point, on Line 19, would be MZ_1, or using the notation of Listing 8.1, MZ[1].

In Line 22 we check the bit of the mantissa just to the left of the binary point. (It should be 1, if the mantissa is normal.) If it is zero, indicated by the expression !M[0], we have a result which is less than one, and we shift the mantissa left, in Lines 23, and 24, until a one shows up in this position. It should be noted that, although it appears that we are both shifting right, and then left, in reality, only one of these two shifts is performed; either we shift right in Lines 19 through 21, and Lines 22 through 24 are skipped, or vice versa.

In Line 25, we finish the algorithm by rounding the result. To simplify the rounding, we assume that we are using RZ, meaning that we just truncate, by setting the round and guard bits to 0.[9]

8.2.4.7 Floating-Point Multiplication

In this section we examine floating-point multiplication. Floating-point division is very similar to floating-point multiplication, and so it is left as an exercise. We start our examination of floating-point multiplication by presenting a sequence of steps, as we did for floating-point addition, based on our knowledge of how to multiply numbers in scientific notation.

We will use the same operands as we did for addition, and as in addition, assume that the operands have been unpacked into the same six registers. So, the problem we will work through is $+1.0111 \times 2^3 \times -1.1011 \times 2^2$. For this example, we will use RM as our rounding method.

1. Add the exponents. The two exponents are added, to arrive at a preliminary exponent for the result. For our example, after this step, the result would be

$$+1.0111 \times -1.1011 \times 2^5.$$

---

[9] If RN were used, you would need to check M[–5], the round bit. If it were a 1, you would then zero out the round, and guard bits, bits M[–5..–6], and then add the constant 0.0001,00 to MZ. (You should be able to work out the code for this.)

2. Multiply the mantissas. The mantissas of the multiplicand and the multiplier are multiplied, using an unsigned integer multiplier. For our example we would perform the following calculation.[10]

$$
\begin{array}{r}
1.0111 \\
\times 1.1011 \\
\hline
1\ 0111 \\
10\ 111 \\
000\ 00 \\
1101\ 1 \\
1\ 1011 \\
\hline
10.11001101
\end{array} \qquad (8.18)
$$

From Calculation 8.18[11] we get the following result.

$$+1 \times -1 \times 10.1100, 11 \times 2^5$$

This notation shows the signs separated out from the multiplication. It still remains to determine the sign of the result.

3. Calculate the sign. We now calculate the sign of the result, using the signs of the operands. After this step, our example is transformed into the calculation

$$-10.1100, 11 \times 2^5.$$

4. Normalize the result. The result must be shifted right, until it is in normal form, and, simultaneously, the exponent must be adjusted. This process yields the following for our example.

$$-1.0110, 01 \times 2^6$$

5. Round the result. We would use one of the four rounding methods to round the result to the floating-point precision being used. As in floating-point addition, after rounding, we might have to re-normalize the result. For our example, using RM, since the result is negative, we would round the mantissa up.

$$-1.0111 \times 2^6$$

---

[10]This multiplication is done in the same way we would multiply two real numbers in decimal; by ignoring the point, multiplying them as integers, and then calculating the location of the point.

[11]Notice that in some columns we get large carry-out values when adding. For example, in Column 5 we get a sum of 4 (100 in binary). For this sum, the lowest digit would be the sum, and the carry-out would be 10 binary (2 in decimal).

```
EZ = EA + EB                               1
MZ = MA * MB                               2
SZ = SA ^ SB                               3
if MZ[1] then                              4
    MZ = MZ >> 1                           5
    EZ = EZ + 1                            6
if SZ then                                 7
    if MZ[-5] | MZ[-6] then                8
        MZ[1..-4] = MZ[1..-4] + [000001]   9
MZ[-5..-6] = [00]                          10
if MZ[1] then                              11
    MZ = MZ >> 1                           12
    EZ = EZ + 1                            13
```

Listing 8.2 Pseudo-code for floating-point multiplier.

We now convert these steps into a more formal algorithm, that can be, in turn, converted into circuitry. The resulting code is displayed in Listing 8.2. Starting at the top of the code, in Line 1 we add the exponents, in Line 2 we multiply the mantissas, and in Line 3 we calculate the sign for Z. The sign is calculated by taking the XOR (the ^ operator) of the sign bits of the operands. This sets the result sign bit to 1, if the signs of the operands are not equal, and sets the result sign to 0, if the operands have the same sign.

In Lines 4 through 6 we normalize the result, in the same way that we did for floating-point addition. Lines 7 through 10 perform the RM round on the result. In Line 7, we first check to see if the result is negative, in which case it may require rounding up. If Z is negative, in Line 8 we check to see if the round, or guard bit is 1. If so, then, in Line 9, we add one to bits 1 through –4 of the result (everything but the round, and guard bits). In Line 10, which is executed whether or not a round up is performed, we truncate off the round and guard bits by zeroing them out. Finally, after rounding, we check the result to see whether we need to re-normalize, in Line 11, and if so, perform the normalization in Lines 12 and 13.

8.3 SUMMARY

In this chapter we have covered the basics of computer arithmetic. We took a look at integer arithmetic and floating-point arithmetic. One point, you may have observed, is that when we get away from just simple integer addition, our computational units start becoming multi-clock-cycle sequential machines. And, in fact, we saw that multiplication takes much more time to perform, than addition. Floating-point operations take even long than integer multiplication. Because arithmetic is such a significant part of what the processor does, designers have long been interested in how arithmetic operations might be sped up.

There are several design ideas that have been implemented to help speed up multiplication, which is one of the big bottlenecks to arithmetic speed. Before we end this chapter, it would be worth mentioning some of the more popular improvements for multiplication.

8.3.1 Wallace Trees

The add-shift multiplier is designed as an algorithm with a loop. In the loop, shifted copies of the multiplicand are added to the product. The way the multiplier is structured, we could, and probably would, use the same adder to perform each addition. The fact that we use the same adder means that the additions must be done in sequence. One improvement that can be made to the multiplier is to do all additions simultaneously. This can be done if we use multiple adders. We might use one adder for each row in which we add the multiplicand to the product. The output of each adder is fed into the next adder, to have the next row added in. This, potentially, would allow us to do the whole computation in one clock-cycle, although because of propagation delay as a row waits for results from other rows, that cycle would have to be fairly long.

Using multiple adders is the idea behind *Wallace tree* multipliers. Wallace trees are a collection of adders, which are connected in a tree structure, and which perform all additions for a multiplication, in a single clock cycle.

8.3.2 ROM Lookup Tables

If we keep thinking about how to improve the speed of multiplication, we might ask whether it is possible to do a multiplication without even doing addition. The answer to this question is yes, it is possible using a *multiplication table*. Instead of multiplying two operands, we might, simply, look our result up in a two-dimensional ROM table, structured very similarly to the multiplication tables we used when learning multiplication.

The problem with doing integer multiplication using only a multiplication table is that the table would have to be very large, and as a large circuit, the table would create latency problems of its own. For example, if the integer word is 32 bits wide, the multiplication table would have to be a

$2^{32} \times 2^{32} = 2^{64} > 10^{18}$ word table to contain results for all possible multiplications. This is unfeasible, and so, multiplication tables are typically used only to multiply small numbers, and larger numbers are multiplied using the full-fledged multiplier. For our example, we might use a table to multiply numbers up to 2^8, resulting in a table size of $2^8 \times 2^8 = 2^{16} = 64K$, which is a much more reasonable size. With this hybrid multiplier, we speed up the smaller, more common multiplications, significantly impacting our overall performance.

8.3.3 Arithmetic Pipelines

Another characteristic of the multiplier that we might examine goes back to the sequential nature of the circuit. The strategy of the Wallace tree is to collapse this sequential structure into a single parallel computation, with many adders. Another approach is to keep the structure sequential, but with many adders; maybe use separate adders to do each iteration of the loop. This might seem wasteful. For instance, the adder used in the first iteration of the add-shift loop would do its addition on the first iteration, and then, for the rest of the calculation, it would sit idle.

While initially, the above architectural proposal for the multiplier looks inefficient, it turns out that it is excellent, if you start thinking about executing several multiplications, simultaneously. If you reconsider the adder used to do the addition of the first iteration of the add-shift loop, in the new scenario, the adder does the first add. But, instead of now sitting idle until the current multiplication is completed, we immediately start the next multiplication operation. Now, we have the first multiplication doing the addition for the second iteration of the loop, using the second iteration adder, and the new multiplication doing its first iteration addition, using the first iteration adder. As the multiplication progresses, we see the two multiplications using distinct adders, simultaneously. This type of processing, where two or more processes are being performed simultaneously, but at different stages of completion, is commonly referred to as *pipelining*, and the type of pipeline we are discussing is referred to as an *arithmetic pipeline*. Pipelining has other uses. We discuss what is called *instruction pipelining* in Section 10.1.2. That discussion should also help clarify arithmetic pipelines.

8.4   EXERCISES

8.1   Do the following calculations using 8-bit binary addition. For each problem indicate if the calculations result in (1) unsigned overflow, (2) signed overflow, or (3) neither.

    a. $123 + 54$

    b. $115 - 124$

c. $92 + 108$

d. $103 - 77$

8.2 You are to multiply 5×11. Develop a table, similar to Table 8.1, that shows the steps performed by the multiplier, for this problem.

8.3 You are to divide $124 \div 10$. Develop a table, similar to Table 8.2, that shows the steps performed by the divider, for this problem.

8.4 Write a BRIM program that reads in a number n, and outputs the value $13n$. Your program should do so, using an add-shift scheme.[12] Your program should have no loops.

8.5 Convert the following decimal numbers into IEEE single precision format. Show the intermediate steps: binary expansion and binary scientific notation.

a. -0.02

b. $+22.40625$

c. $+1.46484375$

8.6 Convert the following IEEE single precision floating-point values to decimal. Show the intermediate steps: binary scientific notation and binary expansion.

a. 1,01111111,1101 1011 1000 0000 0000 000

b. 0,10000111,0110 1101 1011 0110 0000 000

8.7 Show how you would do the addition problem, $-1.1111 \times 2^{-2} + 1.1101 \times 2^{-1}$. Go through the steps from Section 8.2.4.6, and show the state of the problem at each step. Use RN to round.

8.8 Show how you would do the multiplication problem $-1.0101 \times 2^5 \times -1.1101 \times 2^{-2}$. Go through the steps from Section 8.2.4.7, and show the state of the problem at each step. Use RP to round.

8.9 Write a sequence of steps to perform floating-point division.

a. Use your method to perform the calculation $-1.0110 \times 2^4 \div 1.1100 \times 2^2$. Go through your steps, and show the state of the problem at each step. Use RZ to round.

b. Write an algorithm, similar to Listing 8.2, for your division method.

---

[12] Although the BRIM machine has no shift instructions, remember that the operation $R \leftarrow shl\ R$ is equivalent to the operation $R \leftarrow R + R$.

8.10 Modify the algorithm given in Listing 8.1, so that it handles the non-normalized number zero and the number NaN. (Assume that a value of NaN has an exponent of MAX_EXP.[13])

---

[13]MAX_EXP is a symbol you would use in your pseudo-code. It represents the largest possible exponent. (For IEEE single precision this would be the value 11111111.)

Micro-Programmed CPU Design

CONTENTS

W E DISCUSSED HARDWIRED CONTROL IN CHAPTER 7. In that chapter we explained that there are two common control designs in use: hardwired control and micro-programmed control. We explained that it was best to start by describing hardwired control, and that later we would examine micro-programmed control. In this chapter we present the micro-programmed design of the BRIM machine, after explaining the technology used.

We know that micro-instructions, in RTL, describe digital circuits, by specifying pieces of the circuit data-path and pieces of the control circuitry. As such, micro-instructions are less instructions than declarations. That is to say, they describe the composition of a circuit, rather than describing a procedure used in a calculation. They, however, can be adapted to describe procedures, through the use of a sequencer, as described in Section 5.3. When we build these algorithmic descriptions, it is not too much of a stretch to think of a micro-instruction as a statement in an algorithmic language, like $C++$. For instance, the micro instruction

$$R \leftarrow R + S$$

looks a great deal like the $C++$ assignment statement

```
R = R + S;.
```

As a more complex example, the pair of micro-instructions

$$c : R \leftarrow R + 1$$
$$\bar{c} : R \leftarrow 0$$

describes a type of structure that could easily be mimicked using an *if* statement (and, is so mimicked in Verilog).

```
if(c)
    R = R + 1;
else
    R = 0;
```

So, we might continue with the above line of reasoning, and start thinking of micro-instructions as actual software instructions. The implication of this is that processor control, which remember, is built around a collection of micro-instructions, now is no longer done by a piece of hardware, but instead by a piece of software, called the micro-program. This micro-program is being executed by a small sub-processor, inside the processor, called the *sequencer*.[1]

To summarize our discussion so far, the machine cycle is implemented by executing a sequence of micro-instruction, whether you are using hardwired control or micro-control. In hardwired control, the micro-instructions executed are implemented in circuitry. In micro-control, the micro-instructions are stored as code, in a ROM memory called the *micro-ROM* (μROM). The micro-instructions are fetched from the μROM by the sequencer. A device, which we will call the *micro-op decoder*, then uses the micro-instruction to set control lines for the data-path, and perform the operations.

Figure 9.1 is a block diagram of the sequencer. On the right side of Figure 9.1 we see the μROM. Connected to the address port of the μROM is the

---

[1] We are aware that we have already used the term *sequencer* to refer to the counter that produces timing steps in the hardwired control unit. We are simply expanding the definition of a sequencer, for the micro-control unit, to a circuit that not only counts out timing steps, but also performs the control operations required at each timing step.

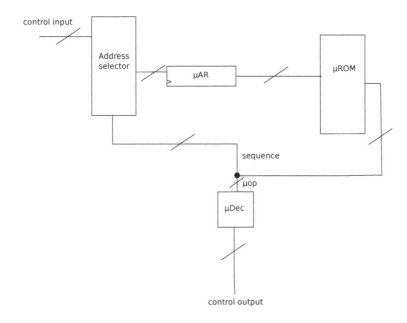

FIGURE 9.1 Structure of the BRIM sequencer.

μAR (micro-address register). This register contains the address of the current micro instruction, and is changed every clock cycle. Given the address in the μAR, the μROM produces a micro-instruction, on its data-out port. We see this instruction split into two fields: the *sequence* field and the *μop* field.

The address of the next micro-instruction to execute is calculated by the *address selector* connected to the input port of the μAR register. As will be described shortly, the address selector uses inputs from the control bus and the sequence field of the micro-instruction to calculate the address of the next instruction.

The final part of the diagram in Figure 9.3 is the *micro-instruction-decoder* (μDec). This component takes the μop field of the micro instruction, and translates it into control signals for the devices on the data-path.

9.1 MICRO-INSTRUCTION FORMAT

We have explained that a micro-instruction will be stored in a μROM unit. We can only store numbers in computer memory, and so, we must translate our RTL instructions into numeric format. The next task will be, then, to describe a usable format for micro-instructions. Since we are using the BRIM processor as our design example, we will tailor our micro-instruction format to the BRIM machine.

When you think about what information must be represented with a micro-instruction like the following,

$$R \leftarrow R + 1$$

you might decide, as we did, to split the micro-instruction into two pieces: a field specifying the micro-operations to be performed, the μop field, and a field specifying the next micro-instruction to be executed, the *sequence* field. We have a lot to say about the μop field, but for now, we need to concentrate on the sequence field.

9.1.1 The Sequence Field

The sequence field specifies how the sequencer can calculate the location of the next instruction to execute. Let us be a little more precise. Suppose that the sequencer is executing a micro-program, stored in the μROM. The micro-program consists of a sequence of micro-instructions, stored one micro-instruction per ROM word. Let us suppose that the sequencer is currently performing the micro-instruction at Address 15. Then, the question is, what instruction to fetch from the μROM next. The sequencer has circuitry available to calculate the address of the next instruction, using several methods.[2]

1. *Increment the current address.* The sequencer would use an adder to add one to the current address, producing the address of the next micro-instruction. In our example, the sequencer would take the Address 15, add one to it, and fetch the next micro-instruction from Address 16.

2. *Unconditional jump to a new address.* The sequencer would use the current address to look up a new address, and this new address would become the address of the next micro-instruction. The new address might be given as a field in the micro-instruction. In our example, the *address* field in the micro-instruction at Address 15 might contain 35, and so the next micro-instruction executed would be fetched from location 35 in the μROM.

3. *Conditional branch to a new address.* The sequencer would use a control input to calculate the new address, if the branch were taken. If the branch were not taken, the current address would be incremented. The branch address calculation might be performed in one of several ways.

   - A hardwired address calculator could be used to calculate the new address.

---

[2]Our list of address calculation methods is not exhaustive. One other possible method might implement the call, and return mechanisms for subroutines.

TABLE 9.1   A simple jump table.

IRX_{15-12}	IRX_{11}	ZX	Address
0000	X	X	01101
0101	1	X	10011
1010	X	1	01111

- The new address might be given as part of the micro-instruction, in the same way as it is for an unconditional branch.

- The new address might be looked up in a ROM table, often called a *jump table*, using the control input to the sequencer as an index into the table.

For our example, let us assume that the new address is looked up in a ROM jump table. Maybe the control input to the sequencer is just the machine language op-code of the instruction being worked on. Let us assume that the op-code specified is 14. Suppose that the fourteenth entry in the jump table is 36. This indicates that the next micro-instruction to be executed is located at location 36 in the μROM.

To choose which of the above methods to employ, the sequence field of the micro-instruction is consulted. The bits of the sequence field store a code indicating the method used to calculate the next micro-instruction.

You might wonder what a jump table looks like. We give a simple example in Table 9.1. For this simple machine, which is based on the BRIM machine, the control inputs to the sequencer are the bits of the IR register, and the status flag Z. Remember that we call these control inputs IRX and ZX, respectively. The combined bits of IRX and ZX are used as an index into the jump table that is shown. Starting with the first row of Table 9.1, if the op-code specified by IRX_{15-12} is zero, then the next micro-instruction would be found at location 01101 (13 in decimal) in the μ-ROM. For the second row of the table, if the op-code were 0101 (5 in decimal), and the addressing mode bit, IRX_{11}, were 1, then the next micro-instruction would be at address 10011 (19 in decimal). And, finally, if the op-code were 1010 (10 in decimal), and the Z flag were set, the next micro-instruction would be fetched from location 01111 (15 in decimal).

9.1.2   The *Select* and *Address* Subfields, and the Address Selector

Continuing with our description of the BRIM sequence field, the micro-format for the BRIM machine is shown in Figure 9.2, with a 28-bit micro-word. The sequence field has been split into two sub-fields: the *select* field and the *address* field. The select field indicates which method is used for calculating the location of the next micro-instruction. We have implemented only two of the addressing methods: unconditional jump, Method 2, which is indicated

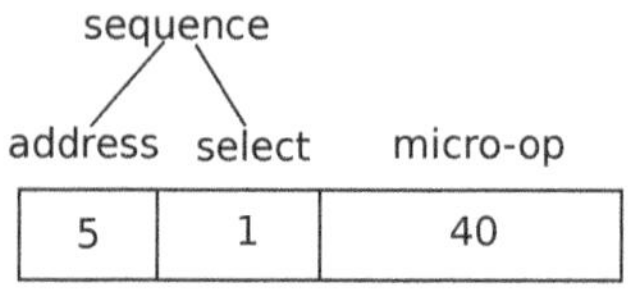

sequence

address select   micro-op

5	1	40

FIGURE 9.2 Format for the BRIM micro-instruction.

by a *select* field value of 0, and conditional branch, Method 3, that uses a jump table, which is indicated by a *select* field value of 1. For Method 2, the target address of the jump is supplied as the value of the *address* field. So for example, if you wished to jump to address 13, the address field would have the 5-bit value 01101 (13 in decimal), and the *select* field would have the value 0, indicating Method 2. The full sequence field would then be the 6-bit value 011010. For conditional branches, the address field is ignored, and the target address is looked up in the jump table. So, the sequence value 000001 would indicate a conditional branch, with 00000 as a dummy target address, and one as the value of the select field.

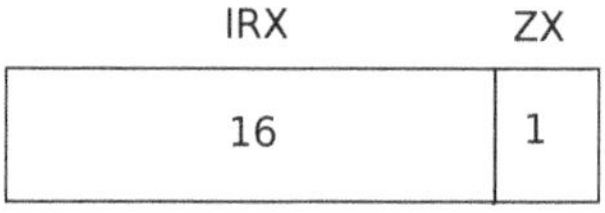

IRX   ZX

16	1

FIGURE 9.3 Format for the BRIM sequencer control word.

As seen in Figure 9.1, the sequence field is one of the inputs to the address selector. The other inputs are control inputs from the bus. We will be referring to this collection of control input bits as the sequencer *control word*. The control word for the BRIM machine is identical to the control word for the simple example of Table 9.1, although the jump table for the BRIM machine is more complex. The format for the control word on the BRIM machine is shown in Figure 9.3. It consists of a collection of seventeen bits: sixteen bits for the IRX inputs, and a single bit for the ZX input.

Now that we understand the sequence field and the control word, we can describe the device that uses them to calculate the address of the next micro-instruction to execute: the address selector. Figure 9.4 shows the structure of the BRIM address selector. The address of the next micro-instruction is calculated by the output MUX. The 0 option of the MUX implements an

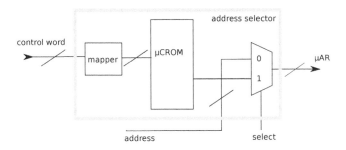

FIGURE 9.4 Structure of the BRIM address selector.

unconditional jump, and the 1 option implements a conditional branch. The MUX is controlled by the select field of the current micro-instruction. You will notice that for an unconditional jump, we simply feed the address field of the current micro-instruction through the MUX.

For a conditional branch, the address of the next micro-instruction is calculated by two components: the *micro-control-ROM* (μCROM), which contains the jump table, and the *mapper*. The mapper is a circuit that is used to form an address out of the control word. In our design of the BRIM machine, not all bits of the control word are used to index the μCROM; only the leftmost five bits of IRX, and the 1-bit ZX value. So, for the BRIM machine, the mapper is rather simple; it is a circuit that simply truncates the 17 input bits, down to 6. It is, of course, quite possible to have a mapper that is much more complex.

Once the mapper has produced the jump table address, and asserted it on the address port of the μCROM, the address of the next micro-instruction is output by the μCROM, and sent through the output MUX.

9.2 MICRO-ARCHITECTURES

We have now described most of the components in Figure 9.1. The final part of the diagram is the *micro-instruction-decoder* (μDec). This component takes the μop field of the micro instruction, and translates it into control signals, for the devices on the data-path. It is difficult to describe, exactly, how the μDec works at this point in time, because we have not even discussed what the μop field looks like, let alone how it is translated. So, our next task is to describe the μop format.

There are, in fact, several ways to design the μop format. They can usually be classified as falling into one of three design types: direct control, horizontal control, and vertical control. The differences between these different control

TABLE 9.2   Micro-operations for Example 9.1.

Code Name	Micro-Inst.	Control
Reg1	$AR \leftarrow TP$	ARLD, STP
Mem1	$X \leftarrow M[AR]$	XLD, ME
Reg2	$AR \leftarrow AR + 1$	INAR
Mem2	$Y \leftarrow M[AR]$	YLD, ME
Reg3	$TP \leftarrow TP + 1$	INTP
Mem3	$M[AR] \leftarrow X + Y$	MW, SADD

designs are probably best explained through an example. For our example, we will be designing a very simple machine that has the capability of executing only one machine-level operation:

$$M[TP + 1] \leftarrow M[TP] + M[TP + 1], TP \leftarrow TP + 1 \qquad (9.1)$$

If you recall our description of the stack machine, from Section 6.4.4, this instruction essentially performs an addition, in the way a stack machine would perform it, by adding the top two elements on a stack, and replacing them with the answer. The micro-code for performing this operation is as follows. It is left as an exercise for you to design the data-path for the machine.

$AR \leftarrow TP$
$X \leftarrow M[AR], AR \leftarrow AR + 1$
$Y \leftarrow M[AR], TP \leftarrow TP + 1$
$M[AR] \leftarrow X + Y$

We have worked out the details for the micro-operations that will be required for this machine. These are given in Table 9.2. In the second column of Table 9.2, we give the micro-operations for the machine, and in the third column, we give the control signals which need to be sent to the data-path, in order to perform the micro-operations. The names of the control lines indicate their functions. ARLD, XLD, and YLD control load lines on the specified registers, INAR and INTP control increment lines on the specified registers, STP and SADD control switches that place the contents of the TP register, and the adder, respectively, on the bus, and ME and MW control memory operation lines.

In the first column of Table 9.2, we have given each micro-operation a name. The name consists of two parts. The first part is a category; either *Reg*, indicating that the micro-operation is associated with register manipulation, or *Mem*, indicating that the micro-operation is associated with memory manipulation. Admittedly, the distinction is a bit arbitrary. The important characteristic of the categorization is that we never perform two register micro-operations at the same time, nor do we ever perform two memory micro-operations simultaneously. The second part of the micro-operation name is a code number. This code number simply ensures that each micro-operation, in each category, has a distinct name.

Having finished explaining Table 9.2, we are now ready to begin describing the three design methods for the μop field. These are

- direct control,

- horizontal control, and

- vertical control.

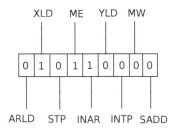

FIGURE 9.5 Direct control μop field.

9.2.1 Direct Control

Direct control is probably the simplest design for the μop field. In direct control, the μop field is a collection of bits, each bit corresponding to a bus control line. For Example 9.1, the μop field might be set up as in Figure 9.5. A bit is allocated in the μop field, for each control line in the data path.

Although we now have a format for the μop field, we have not quite explained how this field is used to perform a micro-instruction. Let us start by examining how a sample micro-instruction would be represented. Take the micro-instruction

$$X \leftarrow M[AR], AR \leftarrow AR + 1. \tag{9.2}$$

This micro-instruction is composed of two micro-operations: one loads the register X from memory, and the other increments the AR register. If we look these operations up in Table 9.2, we find that to do the fetch for the register X, we set the control lines XLD and ME. To increment the AR register, we set the control line INAR. The resulting micro-op field would have the bit configuration shown in Figure 9.5. Notice that we have set the lines XLD, ME, and INAR, and cleared the rest of the lines, to perform only the micro-operations required by the micro-instruction of Example 9.2.

Referring back to Figure 9.1, remember that it is the job of the μDec to convert the μop field specification into control signals for the data path. The reason for previously calling the direct control design "simple" now becomes apparent. Since the μop field is just composed of the control signals themselves, no conversion is necessary, and the μDec is, simply, a straight line connection between the μop field, and the corresponding data-path control lines. This simplicity is a key advantage to direct control.

9.2.2 Horizontal Control

An important advantage of the direct control design is the simplicity of the μDec. The simplicity of this component results in rapid decoding, and execution of micro-instructions. There is also, however, a disadvantage to the direct control design. This disadvantage is not, perhaps, all that visible for the machine of Example 9.1, which only has a few control lines, but when we start dealing with more realistic machines, and even the BRIM machine, the number of control lines increases. Since each control line has a bit in the μop field, the μop field size also grows significantly. This results in fairly long micro-instructions, which require larger μROM units to store them. In addition, for a given micro-instruction, most of the bits in the μop field are zero, which is wasteful of space. To address these problems, we might consider introducing an encoding scheme for the μop field of the direct control design. This is what we do in horizontal control design.

Rather than breaking a micro-operation down into its component control signals, as we do in the direct control design, in the horizontal design we store information in the μop field at the micro-operation level. It is then the job of the μDec to deduce what control is required for the specified micro-operation.

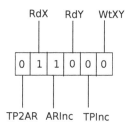

FIGURE 9.6 Horizontal control μop field.

TABLE 9.3   Horizontal fields for
Example 9.1.

Bit Name	Micro-Op.
TP2AR	$AR \leftarrow TP$
RdX	$X \leftarrow M[AR]$
ARInc	$AR \leftarrow AR + 1$
RdY	$Y \leftarrow M[AR]$
TPInc	$TP \leftarrow TP + 1$
WtXY	$M[AR] \leftarrow X + Y$

The information stored in the μop field, for horizontal control, is a collection of bits, indicating the micro-operations being performed for that particular micro-instruction. Figure 9.6 shows a possible format for the μop field, for the machine of Example 9.1. The bits in Figure 9.6 are defined in Table 9.3. Table 9.3 gives the bit name, and the corresponding micro-operation it is specifying. As can be seen, the format given in Figure 9.6 simply contains a bit for every possible micro-operation on the machine for Example 9.1.

For the micro-instruction of Example 9.2, in the μop field, we would indicate that we wish to read a value into the register X, and increment the AR register. This is done by setting the bits RdX and $ARInc$, and leaving all other bits of the μop field as zero, as is done in Figure 9.6.

The horizontal control design, in general, results in a shorter micro-instruction format than the direct control design. Just in our example, we see that the size of the μop field has shrunk from nine bits, for the direct control design, down to six bits, for the horizontal control design. The smaller size of the μop field means a smaller size for the full micro-instruction, which, in turn, means a smaller μROM. While this is a significant advantage, it does not come without a price. In particular, the μDec is no longer a simple array of direct connections. We now require a more complex combinational circuit to translate from micro-operation trigger lines, to control signals.

If we examine the μDec more carefully, it is a combinational circuit, which has the interface shown in Figure 9.7. The output equations for the μDec can be derived by using both Table 9.2 and Table 9.3. The resulting equations are given in Table 9.4. Table 9.4 gives the outputs of the μDec in the first column, as a Boolean definition involving the inputs to the μDec, in the second column. These equations allow us to build the μDec, using the usual technique.

To demonstrate how Table 9.4 was derived, we use ME as an example. We first examined the third column of Table 9.2, for rows which set the signal ME. We find these for micro-operations $X \leftarrow M[AR]$, and $Y \leftarrow M[AR]$.

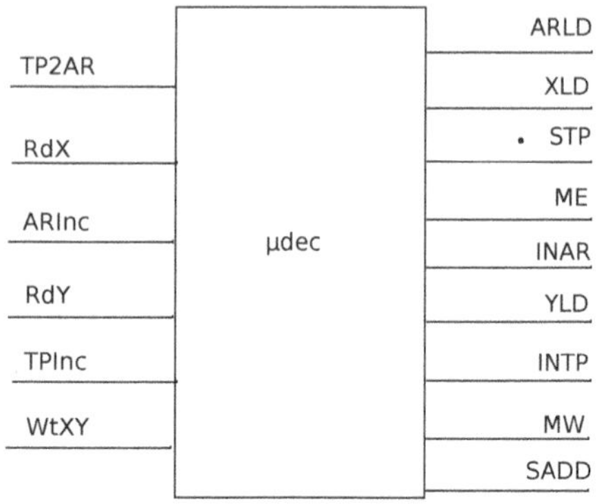

FIGURE 9.7 Interface for the horizontal μDec.

TABLE 9.4 μDec
equations for horizontal
control.

Output	Inputs
ARLD	$TP2AR$
STP	$TP2AR$
XLD	RdX
ME	$RdX + RdY$
INAR	$ARInc$
YLD	RdY
INTP	$TPInc$
MW	$WtXY$
SADD	$WtXY$

We then look these micro-operations up in Table 9.3, and find that they are triggered by the lines RdX and RdY, respectively. This means that the signal ME should be set when signal RdX is set, or when signal RdY is set, resulting in the equation given in Table 9.4.

9.2.3 Vertical Control

When moving from direct control to horizontal control we traded just a small increase in the complexity of the μDec, for a substantial decrease in the size of the μROM. Noticing this improvement, you might ask whether it is possible to do better, by continuing to move in the same direction. That is to say, can we decrease the size of the μop field even further, without an explosion in the complexity of the μDec? Well, the answer is, yes, we can, and we do it by moving to vertical control.

In horizontal control, the μop field is composed of trigger bits, representing the micro-operations of the machine. A fairly straightforward way of reducing the number of bits in the μop field is to encode these trigger bits. That is to say, instead of storing individual trigger bits, we give each trigger signal a number, or code, and instead, store the trigger signal code.

For our example machine, we can number each micro-operation in Table 9.4. These numbers, or codes, will represent the corresponding micro-operations. Because a micro-instruction may perform several micro-operations, to represent a micro-instruction requires more than a single code. A micro-instruction will have to be represented with a set of codes, one for each micro-operation in the micro-instruction. Notice that we do not have to be able to specify all sets of codes; only those that correspond to actual micro-instructions on the machine. So, now we need to discuss how we specify only useful sets of micro-operation codes.

In vertical control design, we typically first divide the micro-operations into groups. These groups are often associated with processor devices. So, for example, we might have a group of micro-operations that work on memory. Then, we might have another group that work with the ALSU. A third group of micro-operations might do register transfers, and so on. How the grouping is done, however, is mostly arbitrary. There is, however, one important property that a group must have: for any two micro-operations in the same group, the two micro-operations are never performed in the same micro-instruction. So, for instance, if you have a memory-read micro-operation, associated with the memory unit, and a memory-write micro-operation, also in the memory group, you would not have a micro-instruction requiring both a memory read and a memory write, simultaneously. This example indicates why we normally choose groups associated with data-path components; most data path components can perform only one operation at a time, and so there is almost no chance of two micro-operations, associated with the component, being required to perform simultaneous actions.

For the machine of Example 9.1, we have split micro-operations into two groups:

- Micro-operations associated with register manipulation.

- Micro-operations associated with memory manipulation.

We have verified that no two micro-operations in the same group are performed in the same micro-instruction. In Table 9.2, as already discussed, the first part of the code name, in the first column, indicates the group to which the micro-operation belongs: memory micro-operations or register micro-operations. Inside a group, we number the micro-operations, much as we number machine operations in the machine language op-code. So, for example, the micro-operation $AR \leftarrow AR + 1$ is in the register group, and it is Micro-operation 2 in that group, so its name is *Reg2*. In the same way, the micro-operation $M[AR] \leftarrow X + Y$ is in the memory group, and has been assigned the index 3, so that its code name is *Mem2*. Notice that group indexes start at 1. The index 0, in each group, is reserved to indicate that no micro-operation from that group is being performed.

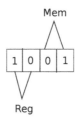

Mem

1 0 0 1

Reg

FIGURE 9.8 Vertical control μop field.

Having settled on a coding scheme for our micro-operations, we are now ready to describe the μop field. The format of the μop field, for the machine of Example 9.1, is shown in Figure 9.8. In the vertical design, each micro-operation group is allocated a field. In Figure 9.8 there are two fields: the 2-bit field to the left is for the register group, and the 2-bit field to the right is for the memory group. Each group field is allocated sufficient bits to store all codes for micro-operations in that group. So, in the machine of Example 9.1, each group has micro-operations numbered from 0 to 3. To represent these codes, we need 2-bit fields, and so, this is the size of the fields in Figure 9.8.

In Figure 9.8 we show values to perform the micro-instruction from Example 9.2. The *Reg* field specifies Micro-operation 2, *Reg2*, which is $AR \leftarrow AR + 1$, and the *Mem* field specifies Micro-operation 1, *Mem1*, which is $X \leftarrow M[AR]$.

Of course, when we encode the micro-operations, we trade an increase in micro-instruction decoding complexity, for a reduction in μROM size. So, the next question is how much more complex is the μDec for vertical control, compared to the μDec for horizontal control.

It is possible to build a vertical control μDec from a horizontal control decoder. The modification is minimal, and is shown in Figure 9.9. We use a horizontal μDec to control the nine data-path control lines. It is fed input by two decoders: the top decoder decodes the *Reg* field of the micro-instruction,

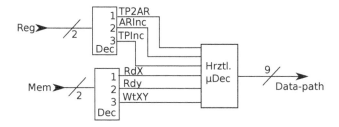

FIGURE 9.9 Vertical μDec.

and the bottom decoder decodes the *Mem* field of the micro-instruction, transforming the two 2-bit field values into the six trigger lines used in horizontal control.

With vertical control, we have shortened the length of the micro-instruction, and, therefore, reduced the size of the μROM. This is done at the expense of a slightly more complex, two-stage μDec.

9.3   MICRO-CONTROL FOR THE BRIM MACHINE

We are now ready to design a micro-programmed control unit for the BRIM processor. To begin the design process, we must choose a control design. We have chosen to implement horizontal control. This choice is a good compromise; decoding the μop field is simpler than with vertical control, and micro-instruction length is shorter than with direct control.

The fact that we are now designing the processor with micro-programmed control changes only the control unit of the BRIM machine. We are still using the same data-path that we did when designing the hardwired control unit, shown in Figure 7.2. The behavior of the processor is still described by the same flowchart used in hardwired design, shown in Figure 7.11. Examining the structure of the BRIM sequencer, shown in Figure 9.1, we see that we only have four pieces that need to be described, in order to complete the micro-programmed design.

- Specify the contents of the μROM.

- Specify the contents of the μCROM.

- Describe the structure of the mapper.

- Give a structural description of the μDec.

TABLE 9.5 BRIM micro-operations.

Micro-Operation	Signal
$AR \leftarrow PC$	PC2AR
$IR \leftarrow M[AR]$	RdIR
$PC \leftarrow PC + 1$	PCInc
$AR \leftarrow IR_{7-0}$	IR2AR
$SR \leftarrow M[AR]$	RdSR
$SR \leftarrow$ if IR_7 then IR_{6-0} else $R[IR_{6-4}]$	SRFtch
$R[IR_{10-8}] \leftarrow SR$	SR2R
$M[AR] \leftarrow R[IR_{10-8}]$	WtR
$DR \leftarrow R[IR_{10-8}]$	R2DR
$R[IR_{10-8}] \leftarrow SR + DR$	RAdd
$Z \leftarrow (SR + DR) = 0$	ZAdd
$R[IR_{10-8}] \leftarrow SR - DR$	RSub
$Z \leftarrow (SR - DR) = 0$	ZSub
$R[IR_{10-8}] \leftarrow SR \wedge DR$	RAnd
$Z \leftarrow (SR \wedge DR) = 0$	ZAnd
$R[IR_{10-8}] \leftarrow SR \vee DR$	ROr
$Z \leftarrow (SR \vee DR) = 0$	ZOr
$R[IR_{10-8}] \leftarrow \overline{DR}$	RNot
$Z \leftarrow \overline{DR} = 0$	ZNot
$PC \leftarrow IR_{7-0}$	IR2PC
$R[IR_{10-8}] \leftarrow D_{in}$	InR
$D_{out} \leftarrow$ if IR_7 then IR_{6-0} else $R[IR_{6-4}]$	OutR

9.3.1 The BRIM Micro-Program

To begin a horizontal control design, we will need a list of micro-operations on the BRIM machine. This information is given in Table 9.5. Table 9.5 was derived from Table 7.2. We simply copied all micro-operations from Table 7.2, and gave them names. Micro-operations involving the sequencer, C, were excluded, since the hardwired sequencer is not part of the micro-programmed machine. The names of the micro-operations become the names of the trigger signals for the horizontal design.

If you count the rows of Table 9.5, you will find that there are 22 micro-operations on the BRIM machine. This means that, for the horizontal control design, the BRIM μop field will have 22 bits. A reasonable format for the field is shown in Figure 9.10. We now have the full specification for the micro-instruction format given in Figure 9.2, and can use this format to convert the micro-instructions from Table 7.2, into numeric form. The results are shown in Tables 9.6 and 9.7.

The first two columns of Tables 9.6 and 9.7 identify the phase of the micro-instruction, taken from Table 7.2, and the location, or address, of the micro-

TABLE 9.6 BRIM micro-program.

Phs	Loc	Add	Se1	PCdAR	PC2IR	IR2AR	RdSR	SR2R	Wt2R	RD2AR	RAdd	RSAdd	RSSub	RAnd	ROr	RNot	R2PC	InR	OutR
F_0	00000	00001	0	1	0	0	0	0	0	0	0	0	0	0	0	0	0	0	0
F_1	00001	00000	1	0	1	0	0	0	0	0	0	0	0	0	0	0	0	0	0
MD_2	00010	00011	0	0	0	1	0	0	0	0	0	0	0	0	0	0	0	0	0
MD_3	00011	00100	0	0	0	0	1	0	0	0	0	0	0	0	0	0	0	0	0
M_4	00100	00000	0	0	0	0	0	0	0	0	0	0	0	0	0	0	0	0	0
MRD_2	00101	00110	0	0	1	0	0	1	0	0	0	0	0	0	0	0	0	0	0
MRD_3	00110	00100	0	0	0	0	1	0	0	0	0	0	0	0	0	0	0	0	0
St_2	00111	01000	0	0	1	0	0	0	1	0	0	0	0	0	0	0	0	0	0
St_3	01000	00000	0	0	0	0	0	0	0	0	0	0	0	0	0	0	0	0	0
Ad_2	01001	01010	0	0	0	0	1	0	0	1	0	0	0	0	0	0	0	0	0
Ad_3	01010	01011	0	0	0	0	0	0	0	0	0	1	0	0	0	0	0	0	0
Ad_4	01011	00000	0	0	0	0	1	0	1	1	0	0	0	0	0	0	0	0	0
Sb_2	01100	01101	0	0	0	0	1	0	0	0	0	0	0	0	0	0	0	0	0
Sb_3	01101	01110	0	0	0	0	0	0	0	0	0	0	0	1	0	0	0	0	0
Sb_4	01110	00000	0	0	0	0	0	0	0	0	0	0	1	0	0	0	0	0	0

TABLE 9.7 BRIM mcro-program (Cont.).

Phs	Loc	Add	Sel1	RCd2AR	PC1nc	IRd2AR	RSFtch	SRR2R	WRt2RDR	RAdd	ZAdd	RSub	ZSub	RAnd	ZAnd	ROr	ZOr	RNot	ZNot	R2PC	InRR	OutR
An_2	01111	10000	0	0	0	0	0	0	1	0	0	0	0	0	0	0	0	0	0	0	0	0
An_3	10000	10001	0	0	0	0	1	0	0	0	0	0	0	0	0	0	0	0	0	0	0	0
An_4	10001	00000	0	0	0	0	0	0	0	0	0	0	0	1	1	0	0	0	0	0	0	0
Or_2	10010	10011	0	0	0	0	0	0	1	0	0	0	0	0	0	0	0	0	0	0	0	0
Or_3	10011	10100	0	0	0	0	1	0	0	0	0	0	0	0	0	0	0	0	0	0	0	0
Or_4	10100	00000	0	0	0	0	0	0	0	0	0	0	0	0	0	1	1	0	0	0	0	0
N_2	10101	10110	0	0	0	0	0	0	1	0	0	0	0	0	0	0	0	0	0	0	0	0
N_3	10110	10111	0	0	0	0	1	0	0	0	0	0	0	0	0	0	0	0	0	0	0	0
N_4	10111	00000	0	0	0	0	0	0	0	0	0	0	0	0	0	0	0	1	1	0	0	0
J_2	11000	00000	0	0	0	0	0	0	0	0	0	0	0	0	0	0	0	0	0	1	0	0
I_2	11001	00000	0	0	0	0	0	0	0	0	0	0	0	0	0	0	0	0	0	0	1	0
Ot_2	11010	00000	0	0	0	0	0	0	0	0	0	0	0	0	0	0	0	0	0	0	0	1

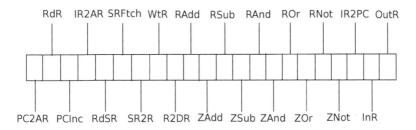

FIGURE 9.10 Horizontal control μop field format for the BRIM machine.

instruction in the μROM. The columns following the first two give the binary micro-instruction. Starting with the third column, and proceeding to the right, we give the address field, the select field, and the bits of the μop field.

The layout of the micro-code is mostly arbitrary, other than the fact that the μROM layout must match the contents of the jump table. (The jump table is still to be specified.) We have chosen to lay out the program in sequential memory words, in the order in which the micro-instructions are listed in Table 7.2.

You should recall that we only have two ways of calculating the location of the next micro-instruction for the BRIM sequencer: unconditional jumps, with a *select* value of 0, and conditional branch, with a *select* value of 1. Places where conditional branches are taken can be determined by examining Figure 7.11. You will observe that the only place where we do conditional branches are from the decision diamonds: Dec, $Is0$, $IsN0$, and Dir. In the phase Dir the branch is conditional on the addressing mode, and remember that this is handled by the mode MUX, and not the CU. So, this leaves the branches from phases Dec, $Is0$, and $IsN0$. Remember that the decision diamonds are incorporated into the action blocks preceding them in the flowchart. We therefore do conditional branches from phase F_1 only. For all other phases we can do an unconditional jump to the next micro-instruction in the sequence. If you look at the row for phase F_1, in Table 9.6, you will see that the *select* field has a value of 1, indicating a conditional branch. The address field value is unused for a conditional branch, and so for the phase F_1 it has been set to zero.

Let us examine another row of Table 9.6. In the row for phase Ad_3, we see that the select field has a value of 0, indicating an unconditional jump. The address that is being jumped to is 01011, the μROM address of the micro-instruction for phase Ad_4. You can verify, on your own, that all jumps in

Tables 9.6 and 9.7 correspond to same jumps indicated in the Figure 7.11 flowchart.

We now examine the μop fields in Tables 9.6, and 9.7. The micro-instruction for phase Ad_3 is

$$SR \leftarrow \text{if } IR_3 \text{ then } IR_{2-0} \text{ else } R[IR_{2-0}], C \leftarrow C + 1.$$

Remember that we ignore all micro-operations on the sequencer counter, C, and so the remaining operation is the operation called $SRFetch$. If you examine Table 9.6, you will see that, for the micro-instruction Ad_3, we have set the $SRFetch$ bit to one, and all other bits of the μop field to zero, performing only the required operation.

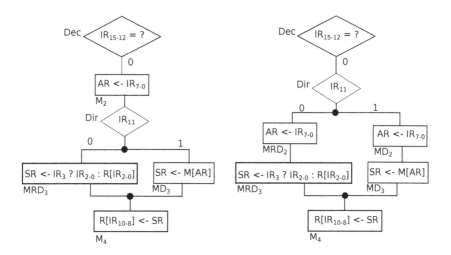

FIGURE 9.11 Changes to the *mov* instruction for micro-control.

There are some slight differences between the phases in Table 7.2, and those in Tables 9.6 and 9.7.

1. The phase M_2 has been replaced by two identical micro-instructions: MD_2, and MRD_2.

2. The phases Bz_2, $Bz_2 \vee \overline{Bz_2}$, Bn_2, $Bn_2 \vee \overline{Bn_2}$, and J_2 have all been compressed into a single phase, J_2.

Difference 1 has been made to simplify the jump table. On the left side of Figure 9.11, we have copied out the pertinent part of Figure 7.11, to explain the effect on the jump table. You can see, in Figure 9.11, that when executing a move machine instruction, we first jump to phase M_2, based on the op-code, and then jump to either MR_3 or MRD_3, based on the addressing mode bit. From here we finish with phase M_4. The two-stage jump process, shown on the left side of Figure 9.11, complicates the use of the jump table, and so to eliminate the two-stage jump, we have, instead, jumped straight into an MD sequence, or an MRD sequence, and placed copies of the phase M_2 in each of these sequences, to perform this micro-operation after the second stage jump, rather than before the jump. This is shown on the right side of Figure 9.11.

Difference 2 is based on the observation that all branch and jump micro-instructions consist of the same micro-operation: IR2PC. What makes the machine instructions *bz*, *bnz*, and *jump* different is not the micro-operation performed, but rather under what conditions the micro-operation is performed. As it turns out, the control for these machine instructions, therefore, is best done, not with the micro-code, but with the jump table. When we presently describe the jump-table, you will hopefully understand this statement. But, for now, we have only one phase that performs a jump, J_2, which is used by all of the branch and jump machine instructions.

With an understanding of the above differences, it should be possible to fully understand Tables 9.6 and 9.7. You can verify, on your own, that all micro-instructions from Table 7.2 have been converted to the correct horizontal control in these two tables.

9.3.2 The BRIM Jump-Table and Mapper

We now turn our attention to the sequencing of the micro-program and the μCROM, or jump-table, and the mapper. The mapper and μCROM are shown in Figure 9.4. Remember that the mapper calculates an address from the control word, and feeds it into the jump-table.

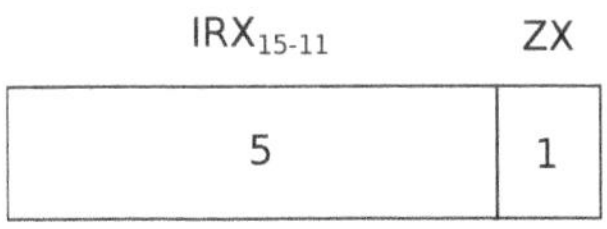

IRX$_{15\text{-}11}$	ZX
5	1

FIGURE 9.12 μCROM address format.

The form of the control word for the BRIM machine is shown in Figure 9.3. It is composed of 17 bits: the sixteen bits of the IR register, and the 1-bit value of the Z flag. We could use this whole 17-bit value as the address to the μCROM, but then we are envisioning a rather large jump-table: a 128K word

TABLE 9.8 Contents of the μCROM.

IRX_{15-12}	IRX_{11}	ZX	Add	Phase
0000	1	X	00010	MD_2
0000	0	X	00101	MRD_2
0001	X	X	00111	St_2
0010	X	X	01001	Ad_2
0011	X	X	01100	Sb_2
0100	X	X	01111	An_2
0101	X	X	10010	Or_2
0110	X	X	10101	N_2
0111	X	1	11000	J_2
0111	X	0	00000	F_0
1000	X	0	11000	J_2
1000	X	1	00000	F_0
1001	X	X	11000	J_2
1010	X	X	11001	I_2
1011	X	X	11010	Ot_2
11XX	X	X	XXXXX	

table. We can get by with a much smaller table, and so the mapper inputs the seventeen bits, and cuts them down to only the six shown in Figure 9.12. This address describes a jump-table of size $2^6 = 64$, which is a much more reasonable size, although still fairly large.[3] The address is composed of the machine instruction op-code, the destination operand address mode bit, and the value of the Z flag.

Let us now consider the contents of the jump-table itself. This content is shown in Table 9.8. In Table 9.8, we have abbreviated a 64-row truth table into sixteen rows, using don't cares. Each row shows an address, on the left side of the table, and the contents on the right side. The rightmost column is given for convenience, and gives the jump target address as a phase.

Table 9.8 was developed using Figure 7.11, the modification in Figure 9.11, and Tables 9.6 and 9.7. In Figure 7.11, we see that the only phase in which we do a conditional branch is F_1. This is not a surprise; the BRIM machine was designed this way, with simplicity in mind. The branches in phase F_1 are made based on the op-code, the destination address mode bit, and the Z flag. Let us do an example to show how a row of Table 9.8 would be constructed.

---

[3]It is possible to decrease the size of the μCROM even further, if we use a more sophisticated mapper. We shall see that there are only fifteen jump target addresses, and so we could use a minimum length jump-table of length 15, if desired, at the expense of more complex mapper circuitry.

Consider the branch from F_1 to Sb_2, in Figure 7.11. The figure indicates that the op-code causing this branch is 3 (0011 in binary). This branch is not dependent on the address mode bit, nor the Z flag. So, this branch results in a row in the jump-table for $IRX_{15-12} = 0011$, $IRX_{11} = X$, and $ZX = X$. The target of the branch is Sb_2, which, from Table 9.6, we find is at address 01100 in the μROM. This results in the fifth row of Table 9.8.

Let us now consider the branch instructions *jump*, *bz*, and *bnz*. To implement these instructions, we jump to either the phase J_2, which loads a new address into the PC, or to F_0, which starts executing the sequentially next machine instruction. For example, for the *bz* machine instruction, we jump from F_1 to J_2, if the op-code is 7 (0111 in binary), and the Z flag is set. This corresponds to the ninth row of Table 9.8. If the op-code is 7, but the Z flag is clear, the corresponding row would be the tenth row. The *bnz* machine operation is handled in a similar fashion, in the thirteenth and fourteenth rows with an inversion of the Z flag. And, the *jump* machine instruction is implemented on the thirteenth row of Table 9.8.

For the *mov* machine instruction, we use the address mode bit of the destination, in the first and second rows of Table 9.8, to determine if we jump to MD_2, for a destination operand that is in direct mode, or if we jump to MRD_2, for a destination operand that is in register direct, or immediate mode. If you review the rest of the rows of Table 9.8, you will see that they also correspond to the structure of Figure 7.11.

9.3.3 The μDec

The last piece of Figure 9.1 to specify is the BRIM μDec. Remember that this combinational circuit converts from micro-operation trigger lines, to data-path control signals. As we did for our simple horizontal control example, in Table 9.4, to describe this circuit, we need to build an equation table, giving the data-path control signals, as functions of the micro-operation trigger signals.

Table 9.9 gives the resulting equations for the μDec. There is a great deal going on in this table, and it needs a bit of explanation. Table 9.9 is constructed using several pieces of information, previously presented. Of prime importance are Table 9.5, from which we get a list of the horizontal control lines for the BRIM machine, and their corresponding micro-operations. We also need a list of the data-path control lines. These we get from Table 7.4. The purpose of some of these data-path control signals is obvious from their names, like RLD, which, it is fairly clear, is the load line on the register file R. With some of the signal lines, however, such as SAD, we might need a bit of a reminder, as to what they do. And so you may need to consult the data-path diagram in Figure 7.2.

TABLE 9.9   BRIM μDec control equations.

Output Signal	Formula
RLD	$SR2R + RAdd + RSub + RAnd + ROr + RNot + InR$
MW	WtR
$IOLD$	$OutR$
$PCLD$	$IR2PC$
$DRLD$	$R2DR$
$SRLD$	$RdSR + SRFetch$
$IRLD$	$RdIR$
$ARLD$	$PC2AR + IR2AR$
ZLD	$ZAdd + ZSub + ZAnd + ZOr + ZNot$
RE	$SRFetch + WtR + R2DR + OutR$
ME	$RdIR + RdSR$
IOE	InR
SPC	$PC2AR$
$SALU$	$SR2R + RAdd + RSub + RAnd + ROr + RNot$
SAD	$IR2AR + IR2PC$
SRF	$SR2R + WtR + R2DR + RAdd + RSub + RAnd + ROr$ $+RNot + InR$
SS	$SRFetch + OutR$
SD	$WtR + R2DR$
$PCIN$	$PCInc$
$ALUOP_0$	$RAnd + ZAnd + RSub + ZSub + RNot + ZNot$
$ALUOP_1$	$RSub + ZSub + RAnd + ZAnd + ROr + ZOr$
$ALUOP_2$	$SR2R + RAnd + ZAnd + ROr + ZOr + RNot + ZNot$

We will now demonstrate the technique used to construct Table 9.9, by building equations for a few representative rows of the table. We start with SRLD, which is representative of the load signals. SRLD would need to be triggered when we are changing the value of the SR register. From the standpoint of Table 9.5, this is in micro-operations where we see the register SR on the left-hand side of an RTL instruction. In Table 9.5, we see two such rows that belong to the signal SRFetch and the signal RdSR. The OR of these two signals then becomes the equation for SRLD, in Table 9.9.

A more challenging equation to do is for the signal SRF, which is representative of the select control lines. Examining Figure 7.2, we see that this control line operates the MUX that feeds the register file its address. With further examination of the data-path, we see that when SRF is triggered we use the destination register number, IR_{10-8}, as the address, and when not triggered, we use the source register number, IR_{6-4}, as the address. So, what we need to find in Tables 9.6 and 9.7 are RTL instructions that use the destination register, either on the left-hand or right-hand side. We find WtR, that writes the destination register to memory; $R2DR$, that copies the destination register to the DR register; $RAdd$, $RSub$, $RAnd$, ROr, and $RNot$, that place the result of an ALU operation in the destination register; and InR, that puts an input value in the destination register. Again, we OR all of these signals together to form the equation for SRF in Table 9.9.

As a last example, we examine the equation for $ALUOP_1$. The signal $ALUOP$ is the 3-bit signal sent to the ALSU, telling it which operation to perform. $ALUOP_1$ is the middle bit of $ALUOP$. When developing the equation for this signal, we need only concern ourselves with RTL instructions that use the ALSU. Further we need only concern ourselves with RTL instructions specifying operations with ALU-op-codes that have their middle bit set. To find these ALSU operations, we turn our gaze to Table 7.2. Specifically, we look at the third column of Table 7.2. Any phase that uses the ALSU has a setting in this column for $ALUOP$. We can then focus on the middle bit of the setting, and select the phases that set this middle bit: Sb_4, An_4, and Or_4. All of these phases contain two micro-operations that use the ALSU. For instance, Sb_4 contains the micro-operations we call $RSub$ and $ZSub$, which both use the ALSU. Listing out all micro-operations that set the bit $ALUOP_1$, we get $RSub$, $ZSub$, $RAnd$, $ZAnd$, ROr, and ZOr. Taking the OR of these signals gives us the equation for $ALUOP_1$ in Table 9.9.

9.4 SUMMARY

We have just finished describing a possible micro-programmed architecture for the BRIM processor. In the process, we considered how we might represent micro-operations, as software, and we described several levels of encoding for the micro-instructions: direct control, horizontal control, and vertical control. We also, briefly, discussed how micro-code sequencing might be done. In particular, we discussed how conditional branching might be implemented.

We now have two ways of implementing control for the processor: hardwired control, which was presented in Chapter 7, and which implements control as circuitry, and micro-programmed control, presented in this chapter, and which implements control as code stored in a memory unit. But, one issue we have not addressed in our discussion, is which of these two implementations is "better." We have placed the word "better" in quotes, because determining which of the two implementations is better, depends on your criteria.

To compare hardwire control with micro-programmed control, we can think of at least three criteria that might be used.

- Ease of modification: How easy is it to change the processor, after the initial design?

- Complexity of the processor: How complex is the circuitry in the processor?

- Speed of execution: How fast can we execute a machine instruction?

When making a design decision as to which design to adopt, the designer would, no doubt, consider all three criteria. The decision on whether to use hardwire control or micro-programmed control would then be dependent on how the designer prioritizes the three criteria.

9.4.1 Ease of Processor Modification

When we compare hardwired control to micro-programmed control, in terms of how easy it is to change the architecture of the processor, you first might ask why we would even want to change the processor design. There are several answers to this question.

- We might want to add capability to a processor, and produce a new model of the processor.

- We might be fixing bugs in our original processor design.

- We might want to use the same basic processor, but impose a completely new architecture on it.

If you now accept that these are good reasons to change a processor design, the next issue to consider is what is involved in processor modification. If we wish to modify a hardwired design, the process is a little involved. We first must make changes to the data-path and the behavioral description of the control unit. This work is primarily carried out on design documents. Once the appropriate changes have been made to the design documents, we must then use the design to produce new circuitry.

The important point about modifying hardwired design is that we cannot change processors that have already been built; all we can do is build new processors, with the design changes incorporated. We can contrast this situation with design changes to a micro-programmed processor. To change the design of the micro-programmed processor, we would start by making document changes, as with the hardwired design, and then continue by changing the micro-program to reflect the changes. The interesting difference between hardwired and micro-programmed processors is that when we get to building the processor with micro-programming, we do not need to build a completely new processor, to include the changes. We simply replace the contents of the μROM, and μCROM with new code.

So, what we see is that to change the design of a processor with hardwired control, the process involves a full circuit redesign. With micro-programmed control, the redesign process is basically a matter of software modification. As a consequence, we would normally have to concede that micro-programmed processors are easier to modify than hardwired processors.[4]

9.4.2 Complexity of the Processor Circuitry

If we compare the processor circuitry involved in hardwired design to that involved in micro-programmed design, we might start by looking at the hardwired design. Remember the major components of the processor in hardwired design: a sequencer that drives a rather complex control circuit, which in turn drives the data path. Next, if we look at the processor design for the micro-programmed processor, it is composed of a relatively simple sequencer circuit, although the sequencer is more complex than that for the hardwired control. However, there is a minimal amount of control circuitry in the micro-programmed processor; most control is now specified as part of the instructions in the μROM. This radically reduces the complexity of the micro-programmed circuitry, over the circuitry needed for hardwired control.

9.4.3 Speed of Machine Instruction Execution

Up till now, we have seen advantage biased toward the micro-programmed processor. Yet, many hardwired processors are being built, currently. The

---

[4]It should be said that micro-controlled processors are only easier to modify than hardwired controlled processors, if the modifications are relatively minor. If the modifications are major, changes to the data-path might be required. Once you change the data-path, you probably will have to change things like the micro-instruction format and the bus structure, resulting in massive amounts of circuitry changes, as well as micro-program changes.

TABLE 9.10   Specification for Exercise 9.1.

Assembly Code	Machine Code	Meaning
dbl R_1	1110 0 R_1# 00000000	$R_1 \leftarrow 2 \times R_1$

TABLE 9.11   Specification for Exercise 9.2.

Assembly Code	Machine instruction Code	Meaning
skip	1110 0 000 00000000	$PC \leftarrow PC + 1$

reason is that this last criterion falls in the favor of hardwired control. Let us discuss, briefly, why this is so.

For the hardwired processor, we have a custom circuit, designed and built to execute the instructions of the machine. However, as we already mentioned, the circuitry of the hardwired processor is more complex than that for the micro-programmed processor. Usually more complex circuitry means that the device speed is slower.

For the micro-programmed processor, we have simpler circuitry. However, this is not the only factor that affects the speed of the device. In the micro-controlled processor, every time we execute a machine instruction, we must fetch several micro-instructions from the μROM. These fetches, relative to custom circuitry speed, are usually very slow, and slow down the processor, significantly. The additional circuitry needed to decode the μop field (the μDec) also contributes to the slow-down. The slow-down is usually enough to overcome, and then some, the speed-up caused by the simpler circuitry. The result is that a hardwired processor is, typically, faster than an equivalent micro-programmed processor.

As mentioned before, which design you choose, whether hardwired, or micro-programmed, depends on your priority as a designer. If you are primarily interest in speed, then you might choose a hardwired design. If your major concern is being able to change your design, you might, reasonably, choose the micro-programmed design.

9.5   EXERCISES

9.1   We wish to add the machine instruction *dbl* to the register implicit machine. This instruction doubles the value in a given register. Its specification is given in Table 9.10. Make modifications to the BRIM horizontal control machine to include this instruction. In particular, make the necessary changes to Table 9.5, Tables 9.6 and 9.7, and Table 9.8.

9.2   This exercise is similar to Exercise 9.1. Instead, you are adding the

TABLE 9.12 Specification for Exercise 9.3.

Assembly Code	Machine Code	Meaning
swap R_1, R_2	1110 0 R_1# 0000 R_2#	$R[R_1] \leftarrow R[R_2], R[R_2] \leftarrow R[R_1]$

instruction *skip*, that skips the next instruction. The specifications are given in Table 9.11.

9.3 This exercise is similar to Exercise 9.1. Instead, you are adding the instruction *swap* that swaps the contents of two registers. The specifications are given in Table 9.12. (Assume that the addressing mode for R_2 is always direct mode.)

9.4 Redesign the BRIM sequencer for direct control. Specifically, redraw Tables 9.6 and 9.7, reworking the μop field.

9.5 Redesign the BRIM sequencer for vertical control. Specifically, redraw Tables 9.6 and 9.7, reworking the μop field. Your vertical design should use a μop field, with four sub-fields:

- *mem*: includes the micro-operations RdR, $RdSR$, and WtR.

- *reg*: includes the micro-operations $PC2AR$, $PCInc$, $IR2AR$, $SRFetch$, $SR2R$, $R2DR$, $IR2PC$, InR, and $OutR$.

- *alsu*: includes the micro-operations $RAdd$, $RSub$, $RAnd$, ROr, and $RNot$.

- *status*: includes the micro-operations $ZAdd$, $ZSub$, $ZAnd$, ZOr, and $ZNot$.

9.6 Build the μDec for the machine described in Exercise 9.5. Specifically, draw a decoder diagram, similar to Figure 9.9.

9.7 As a footnote, it is mentioned that one possible sequencing option for a sequencer is the call-return mechanism needed to implement micro-subroutines. In a call sequencing request, the sequencer would do an unconditional jump, but also store the return address in a micro-return-address register (μRA). For a return, the value stored in the μRA would be loaded into the μAR. Assume that the return address would always be the current value of the μAR, plus one. To be more precise, we define the call and return as follows, where Add is the value of the address field in the micro-instruction.

- Call:

$$\mu AR \leftarrow Add, \mu RA \leftarrow \mu AR + 1$$

- Return:

$$\mu AR \leftarrow \mu RA$$

Make modifications to Figure 9.4 to implement the micro-subroutine mechanisms just described.

A Few Last Topics

CONTENTS

IN A BOOK AS CONCENTRATED AS THIS ONE, we unfortunately only have space for the very essential topics in computer organization and architecture. There are, however, some more advanced topics that are extremely interesting, and although we do not have the space to cover them in depth, they are worth a mention. So, in this chapter we take a brief look at a few of these more advanced topics. We give you only enough information to understand the concepts involved in these topics. For a more in-depth discussion, you will have to do a bit of book research. Documents covering these more advanced topics are ubiquitous.

Since the advent of the electronic computer, users have been cursing the limitations of this particular type of device. When we are discussing the computer at the machine level and circuit level, two limitations come to the forefront as particularly significant: the speed at which we can do a calculation, and the amount of storage space we have to work with while performing the calculation. In this chapter we examine several techniques that have been used to ameliorate the problems associated with these two limits. We look at a couple of techniques used to address the issue of calculation speed, and one technique that has done a great deal in solving the problem of storage space.

10.1 DECREASING EXECUTION TIME

The question we now ask is whether there is a technique we can use in processor design, to speed up the execution of a program. The answer to this question is yes, and there are actually several techniques we can use, and that are in common usage. We will focus on two ways of improving performance.

- Decrease the amount of memory access.

- Have the processor work on multiple instructions at one time.

To decrease the amount of memory access, we will use a device called a *cache unit*. To allow the processor to work on several instructions at once, we use a processor design that incorporates an *instruction pipeline*.

10.1.1 Cache Memory

The processor needs to have access to a memory unit. The problem is that accessing memory is slow, compared to the speed at which ALSU operations can be performed. There are several reasons for this. One is that, quite often, the main memory unit that the processor is using is a DRAM unit, which is a technology that has relatively large latency associated with it. Another is that the main memory chip is usually external to the processor, resulting in longer transfer times. On this last point, you could argue that you could simply put the RAM unit inside the processor. But, including such a large number of storage devices in the processor increases the chip area of the processor, and, among other problems, makes doing this difficult.[1]

A common technique used to improve the performance of the data fetches involves placing a small, specialized memory unit, called a *cache* unit, in the processor. The cache must be small enough so that its inclusion in the processor does not significantly degrade the performance of the processor. The cache unit is loaded with data that might be needed in the near future.

---

[1] Although including memory in a processor is problematic, the reverse is feasible, within limits. A C-RAM (*Computational RAM*) is a RAM unit containing a small processing element.

The use of a cache unit can be compared to an analogy. Consider a retail shoe store. The store keeps an inventory of shoe models that it feels are useful. A shoe is useful if it sells well. Keeping an inventory of popular shoes, allows the retailer to quickly sell a shoe, without having to order shoes from the distributer every time a customer wishes to buy a shoe.

For a processor, we keep an inventory of data that most likely will be needed in the near future, in a small RAM unit, often inside the processor. In order to make caching work, this smaller memory unit will need to have fast access, and so it is often a small SRAM unit. We fetch data from this small inventory, and if they are present, we have a situation called a *cache hit*. But, just as with the shoe store, where we may request a shoe that is out of stock, and need to wait for the arrival of a shipment from a distributer, if we request a datum that is not in the cache unit, a situation called a *cache miss*, we may have to wait for a full-fledged memory access.

One question, that might have occurred to you, is how do you tell if a datum is going to be needed soon. Determining if a datum will be used soon might sound complicated, but it turns out that two properties of computer programs allow us to significantly simplify this task. We phrase these two properties as the following rules of thumb.

- *Temporal Locality.* If a particular piece of data has just been executed, it will probably be executed again soon.

- *Spatial Locality.* If a particular piece of data has just been executed, data close to it will probably be executed soon.

These two properties result from the nature of machine code. When the data we are fetching are instructions in a program, the instructions that are executed most are often instructions in loops. So, if an instruction is executed, chances are that it will be executed again soon, the next time the loop is executed, giving us temporal locality. Also, we would guess that instructions in its vicinity would soon be executed, because program instructions are executed sequentially, giving us one type of spatial locality, called *sequential locality*.

There are several caching schemes that take advantage of temporal and spatial locality. The simplest is direct mapped caching. We will briefly examine this scheme.

10.1.1.1 Direct Mapped Cache

The defining characteristic of direct mapped caching is that any given datum is always stored at a fixed location in the cache unit. The best way to explain this opening statement is with an example.

Suppose that we have a DRAM unit of size 64×8, which means that the addresses are 6 bits wide. The processor has a cache unit. The cache unit is a 4×20 SRAM unit. Each of the words of the cache unit is called a *cache entry*. So, the cache unit has four entries, with 2-bit addresses. Each 20-bit entry contains two words of data, fetched from DRAM. When a datum is fetched from the DRAM unit, it, and an adjacent word, are placed into a fixed entry in the cache unit. The two data words occupy 16 bits of the 20-bit word, and the remaining 4 bits are used for bookkeeping, as explained below.

A simplistic way of determining where to place a data word is to take its address modulus 4. We illustrate this with an example. Suppose that we are fetching a data word from location 010110 in the DRAM. We would calculate 010110 mod 4, to find its location in the cache unit. This, in binary, turns out to be nothing more than chopping out the two low-order bits, giving us the cache location of 10. Notice that we use 4 as the modulus, because this is the length of the cache unit, and by so doing, the result will be two bits, the size of the cache address.

You might now observe a problem with this mapping scheme. We just observed that a word at DRAM location 010110 will map to the cache location 10, but so will a word from location 010010. If we access the location 010110, and load its contents into cache, and later access another location with address 010010, the cache entry to which it maps will be occupied. This phenomenon is called a cache *conflict*. When a conflict occurs, the older datum must be replaced by the newer datum.

Notice that a conflict can occur, even if the cache is partially unoccupied. We may be forced to remove a datum that might soon be needed, even though there is room in the cache for both the newer and older datum. This is one of the problems with direct mapped cache.[2]

To elaborate further on what we know so far about direct mapped caching, when the processor is to fetch a piece of data, it first consults the cache unit. If the datum is found in cache, a condition called a *cache hit*, the datum is fetched from cache, and no DRAM access is necessary. If the datum has not been loaded into cache, a condition called a *cache miss*, it is fetched from memory, loaded into cache, and then accessed from cache, by the processor.

---

[2]There are other caching schemes, in common use, that address this problem. In *associative caching*, data can be placed anywhere in cache, which has the effect that there are no conflicts unless the cache is completely full. Another caching scheme, called *set associative caching*, is a hybrid of direct mapping and associative caching. In set associative caching, a data word is direct mapped to a cache entry, but that cache entry, called a *set*, contains slots for multiple words of data. This has the effect that a conflict occurs only if the set is full.

Let us now examine how we determine if a data word is in cache or not. The root question is, when examining a cache entry, how do you know whether or not the cache entry is empty, and if it is determined that the entry is occupied, how do you tell which of the many data words that maps to that entry is present? It becomes fairly apparent that we are going to have to store, not only data in the cache entry, but also some bookkeeping information that will help us identify the contents of the entry.

	tag	V	data 0	1
00	101	1	10100001	00101111
01	000	0	10000001	00011100
10	011	1	11110111	00001000
11	110	1	01100110	10011001

FIGURE 10.1 Direct mapped cache format.

Figure 10.1 shows our example cache, a plausible cache format for direct mapping, and some sample cache content. Each entry, as previously stated, is 20 bits wide. These twenty bits are split into three fields: the *tag* field, the *valid*, or V-bit, and the *data* field. Starting from the right, the data field contains the two data words that have been stored. The two left fields, at this point, are more interesting. These two fields, the tag and valid fields, are the bookkeeping fields, already mentioned, that are needed to identify the contents of the cache entry.

As pointed out, the data field of the cache entry contains two words of data. If you examine Figure 10.1, you will observe two columns in the data field: one marked with the header 0, and the other marked with the header 1. These header values, 0 and 1, are called the *offset*. We store one word of data under offset 0, and another word under offset 1. The two words of data are consecutive in memory.

What is being indicated here is that when we fetch a word of data from DRAM into cache, as previously stated, we actually fetch two words of data at a time. This is an attempt to take increased advantage of spatial locality.

We fetch the word of data we are interested in, and also an adjacent word, with the view that it may be needed soon.

The data field contains two consecutive words of data, only if the cache entry is occupied. The valid field is used to determine if a cache entry is empty or not. If the V-bit is zero, this indicates that the entry is still empty. If the V-bit is one, the entry has valid data loaded into it. Notice that when the processor is powered up, all V-bits have zero as their value. Once data is loaded into the entry, the V-bit for that entry will change to a value of one, and remain so, until power-down. So, we see, from Figure 10.1, that Entry 1 in the cache unit is empty, and all others have valid content.

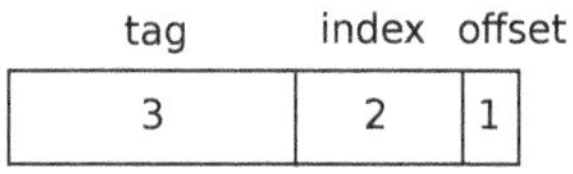

tag	index	offset
3	2	1

FIGURE 10.2 Direct mapped address fields.

To accommodate cache conflicts, and multi-word cache entries, we will have to modify our simple modulus 4 direct mapping scheme. Let us examine, more closely, how we map a word of data, in DRAM, to a cache entry. We use the DRAM address of the word to determine the following: the entry number, or *cache index*, for the datum, the tag for the entry, and the offset for the datum. As an example, consider a data word, 00011100, at DRAM address 000011. We split the address into three fields, as shown in Figure 10.2. For our example, we would find that the offset is the low-order bit of the address, 1, the index is composed of the next two higher bits, 01, and the tag would be the leftmost three bits, 000. To store this word in the cache, we would store it into Entry 1, as indicated by the index value of 01. The datum would be placed into the data slot for Offset 1, and we would record the value 000 as its tag value. This is exactly the situation shown in Figure 10.1, for Entry 1. At the same time as we fetch this word, we also fetch the previous word, 10000001, at address 000010, and store it at Offset 0 of the same cache entry. Notice that addresses of the two fetched words, 000010 and 000011, differ only by their offset. We are fetching data words with both offset values, simultaneously.

The above example illustrates cache placement. We now consider cache search. Suppose that instead of fetching a data word into cache, we are trying to determine if a data word at DRAM location 001111 is already in the cache. We first break the address into the three fields of Figure 10.2; this gives us an offset of 1, an index of 11 (3 in decimal), and a tag of 001. Using the index, we look at Entry 3, and discover that the V-bit is set, and so the cache entry is occupied. We then take the tag, match it against the tag field of the cache entry, and observe that the two do not match; the tag for the data word being accessed is 001, and the tag in the cache entry is 110. (This tells us that the data stored in the cache entry are from addresses 110110 and 110111.) As a consequence, we discover that the datum from DRAM location 001111 is not in the cache, and must then be fetched from DRAM, to replace the current entry.

Of course, caches and main memories come in various sizes. This affects the sizes of the address fields. Let us discuss how address mapping is done in the general case. Suppose that a main memory has a size of $2^m \times n$. This means that the main memory address is m bits wide. The cache can differ from the example of Figure 10.1, in both length and width. When we say that the width may differ, we are talking about the size of the data field; we may decide, instead of storing two words of data in a cache entry, that we may want to store more, or fewer words of data in the entry. In general, assume that our data field stores 2^w words from the main memory. Assume also that we have settled on using a cache unit of length 2^k entries. Knowing the length of the cache, the width of the data field, and the length of the main memory, we can calculate the sizes of the tag, index, and offset fields. Starting with the offset, since each entry stores 2^w main memory words, the offset would have to be w bits. Since the cache has 2^k entries, the index field would be k bits. The tag field would be all of the remaining bits of the address, so it would have $m - k - w$ bits.

Another question you might ask about the cache unit is how many bits wide must each entry be. We can calculate this from the available information, for the entry format shown in Figure 10.1. We know that the data field will have w main memory words in it, with each word being n bits wide, giving us the full field width of $w \cdot n$ bits. The valid bit contributes one bit to the entry, and the tag field, we already know, is $m - k - w$ bits. The total width of a cache entry is $wn + m - k - w + 1$.

tag	index	offset
3	4	3

FIGURE 10.3 Address fields for example.

As an example, to demonstrate our general-case formulas, assume we have a system with the following specifications.

- A DRAM that is of size $1K \times 8$ ($2^{10} \times 8$).

- A cache with eight (2^3) data words per entry, and 16 (2^4) entries.

Then, the 10-bit DRAM address would be divided into fields, as shown in Figure 10.3. The offset would be 3 bits, the index would be 4 bits, and the tag would be $10 - 4 - 3 = 3$ bits. The width of a cache entry would be the sum of the size of the data field, the size of the tag, and the size of the valid bit, or $8 \cdot 8 + (10 - 4 - 3) + 1 = 68$ bits. Therefore, our cache unit is of size 16×68.

10.1.1.2 Writing to Cache

You may have noticed that in the discussion in Section 10.1.1.1 we addressed only the procedure for bringing data from DRAM into the CPU. This discussion only addresses memory read operations.

When the processor gets an instruction to write to a memory location, it looks in the cache to determine what to do. If the address results in a hit, then the processor might write the data to the cache unit. Notice that the changes have not yet been written to main memory, and so the cache and the main memory are inconsistent. If the address reference results in a miss, then the processor might read the main memory, and load the contents into the cache, just as it would for any read operation. Then, the processor would write the changes to the cache, just as when there is a cache hit.

Notice that for both the cache hit and the cache miss, the writes are performed only to the cache unit. At some point in time, the main memory must be updated. So, in our hypothetical system, we might decide that when the cache entry is replaced because of a conflict, before overwriting it, we first write the entry out to the main memory.

To summarize our design, writes are performed to the cache unit only. When a cache entry is swapped out, the entry is written to main memory, if it is has not been modified. We use the term *dirty* to signify a cache entry that has had data written to it. A *clean* entry is one that has only been read, and does not need to be written to memory upon replacement.

The scheme of memory update that we have just described is called the *write-back* scheme. It is not the only scheme used. In the *write-through* scheme, you would write changes to the main memory unit at the same time as you write to the cache unit. This method has the advantage of maintaining consistency between the cache, and the main memory. Its disadvantage is that every time a write is performed, you must access memory, which invalidates, to a degree, the very reason for implementing cache in the first place—to decrease main memory accesses.

10.1.1.3 Cache Performance

Adding a cache unit to a processor will almost assuredly improve the performance of the processor. The question then is only one of how much of an improvement will result. The problem with trying to analyze improvement due to caching is that the improvement is largely dependent on the behavior of the executing machine program. As a consequence, for any analysis you do, you will always be able to find programs that cause the performance of the machine to differ from your results. The best we can hope for is an "average-case" analysis.

To start our discussion of cache performance we will define four terms. First, we define what we will call the *hit ratio*, R_h, and the *miss ratio*, R_m, for a sequence of n memory accesses.

$$R_h = \frac{N_h}{n}, R_m = 1 - R_h \tag{10.1}$$

In Equation 10.1, N_h is the number of cache hits out of the n total accesses. It should be clear that the hit ratio, R_h, is between 0 and 1. The miss ratio, R_m, is then just one minus the hit ratio. This works out to be the number of cache misses, relative to the total number of accesses.

In addition, we define two other terms: the *expected hit rate*, h, and the *expected miss rate*, m.

$$h = E(R_h), m = 1 - h \tag{10.2}$$

The value h is independent of the sequence of memory accesses, and can quite appropriately be thought of as the *a priori* probability of a cache hit. Similarly, the *a priori* probability of a cache miss is m.

With the definitions in Equation 10.2, we can write a formula for the expected time to fetch a datum on a system with a cache unit, which we will call Machine A.

$$T_A = T_h + (1 - h) \cdot T_m \tag{10.3}$$

In this formula T_h is the time to process a cache hit, and T_m is the excess time to process a cache miss, also called the *miss penalty*.[3] The total expected time to process a memory access, T_A, is the time to process a hit, added to the probability of a miss, multiplied by the excess time to process a miss.

We are also going to need a formula for the expected time to fetch a datum on a base machine; a machine with no cache, which we will call Machine B.

$$T_B = T_m \tag{10.4}$$

---

[3]The time to process a hit, T_h, is just the time to access cache. The miss penalty, T_m, includes only the time to access memory.

In other words, the time to fetch a piece of data on a machine with no cache unit is just the time to fetch the instruction from memory.

We can now define performance as a ratio of the fetch times.

$$P_{A,B} = \frac{T_B}{T_A} \qquad (10.5)$$

Let us illustrate how we would use this formula, with an example. Suppose that, on the cache machine, Machine A, the time to fetch an instruction from cache, T_h, is 2 clock cycles, the time to fetch an instruction from main memory, T_m, is 8 clock cycles, and the hit rate, h, is 0.75. Then the performance of Machine A, relative to the base machine, Machine B, is the following.

$$P_{A,B} = \frac{8}{2 + (1 - 0.75) \cdot 8} = \frac{8}{4} = 2 \qquad (10.6)$$

In other words, the fetch time, on a machine with these cache characteristics, is twice as fast as the fetch time on a machine without cache.

10.1.2 Instruction Pipelining

Let's look at another idea for speeding up the execution of a program. In addition to speeding up the execution of each instruction, we might also decide to try increasing the number of instructions we can execute in a given time period. (The number of instructions executed per second is called the *throughput* of a machine.) This is different than speeding up the processor, which decreases the time period to execute a single instruction. The time to execute an instruction remains the same; the number of instructions we can send through the processor in a given time period increases.

If we are to send more instructions through the processor in a certain time period, without decreasing the time needed to execute an instruction, we are probably going to have to execute several instructions at a time. The ability to execute several instructions at a time is called *parallelism*. And, there are several ways to introduce parallelism into a system.

One way to introduce parallelism is to build a computer with several separate processors. So, for instance, you might have a machine with sixteen processors. Conceivably, you would then have a system capable of executing sixteen instructions, simultaneously. But, several interesting issues have to be addressed in order to implement this amazing machine: how do you divvy up the instructions in a program between processors, how do the processors communicate, and how do you organize memory, so that the processors stay out of the way of each other, when accessing memory? These issues, among others, all must be addressed, to make this architecture possible.

Using parallel processors is high on the scale of parallelism. There are several designs that are a little less parallel, but still yield an increase in performance. One scheme is to combine processors in one chip. In this scheme, coupling is possible. Coupled processors share devices, like memory units. A processor composed of several coupled component processors is often referred to as a *multicore processor*. We refer to each component processor as a *core*. The word "core" emphasizes the fact that a core does not have to be a fully autonomous processor.

If you were to take the concept of coupling to its extreme, having cores share more and more devices, the cores would eventually merge into one processor. So, an interesting question is whether it is still possible to implement parallelism with a simple, standard, single data-path, single control unit processor. Yes, is the answer; it is possible for several instructions to share the single data-path, simultaneously. This special processor design is called a *pipelined* processor. The pipelined processor requires some modifications to the normal bus organization, to support pipelining.

If we were to talk about pipelining, but with reference to software rather than hardware, we would probably call it *stream processing*. In stream processing, you split your task up into sub-tasks. These sub-tasks are handled by individual processes, and arranged in a sequence. Each process receives its input from the previous process in the sequence, does its calculations, and feeds its output in, as the input to the next process in the sequence.

When talking about hardware, this stream architecture is referred to as a pipe, or pipeline. To use pipes in hardware, we must reorganize the processor into a stream of circuits, one circuit feeding the next down-stream circuit its input.

Figure 10.4 shows a possible pipe architecture for a simple processor. There are many details that are being neglected in this diagram, but it serves well to help with a discussion that focuses solely on the major concepts of a pipe architecture. At the bottom of Figure 10.4 is a sequence of devices that forms the pipeline. These devices are usually referred to as the *stages* of the pipeline. Each stage gets its input from the previous stage upstream, and feeds its output, as input, to the next stage downstream. The pipe in Figure 10.4 is a 5-stage pipeline, consisting of the stages *Fetch*, *Decode*, *Execute*, *Memory*, and *Write-Back*.

Above the pipeline, in Figure 10.4, are several devices that we have used in the BRIM data-path: a memory unit, a register file, and an ALSU. The pipeline uses these devices to implement the machine cycle.

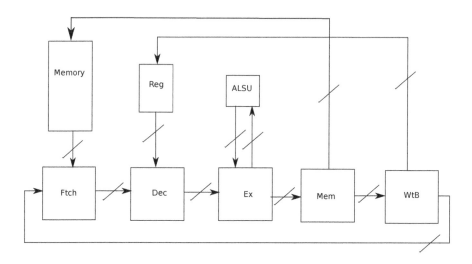

FIGURE 10.4 Processor pipeline stages.

The machine cycle begins with the Fetch Stage. The Fetch Stage circuitry reads an instruction from the memory unit. The resulting instruction is then passed to the Decode Stage. Here the instruction's op-code is decoded. Any operands for the instruction are fetched from the register file. All results are passed to the Execute Stage. In the Execute Stage, if the instruction requires data manipulation, the ALSU is used to operate on the operands, and produce a result. The result is then passed to the Memory Stage. In the Memory Stage any required memory access is made. Results from the Memory Stage are then passed to the Write-Back Stage. In the Write-Back Stage, any results that need to be, are written to the register file. Then the machine cycle begins again at the Fetch Stage.

In the simplest scenario, each stage of the pipe would do its work in one clock cycle. (This is often the case, but not always.) And, the clock cycle must be long enough to accommodate the slowest of the stages. This means that to execute an instruction takes five clock cycles, for a five-stage machine. In general, on a s-stage machine, an instruction is executed in s cycles.[4]

An important characteristic of the stages is that each stage is not *implicitly coupled* with any of the other stages. Coupling refers to a dependency between one stage and another. There are always *explicit couplings* between

---

[4]Notice that every instruction on a 5-stage machine takes five cycles to execute, unlike on the BRIM machine, where different instructions finish in different numbers of cycles. If you think about this, this means that some of the faster instructions are idling, or wasting time on the pipeline machine.

TABLE 10.1 Operation of a five-stage
pipeline.

Cycle	Ftch	Dec	Ex	Mem	WtB
0	I1				
1	I2	I1			
2	I3	I2	I1		
3	I4	I3	I2	I1	
4	I5	I4	I3	I2	I1
5		I5	I4	I3	I2
6			I5	I4	I3
7				I5	I4
8					I5

stages; stages are dependent on each other for their input and output. When we say that the stages are not implicitly coupled, we are saying that, other than for input and output, there are no other dependencies between stages. The practical implication of this is that when two stages, like Fetch and Memory, share a resource, like the memory unit, the resource must be capable of handling requests from both stages, simultaneously, or we would have an implicit coupling, in terms of synchronization. As we mentioned, we are ignoring a lot of detail, but obviously, this memory problem is something that would need to be addressed for the pipeline architecture of Figure 10.4; along with other specialized hardware, we would need a memory unit capable of doing a read and a write operation simultaneously.[5]

If we simply execute one instruction at a time, this property of implicit decoupling is not at all important. But, we have developed pipelines so that we can execute several instructions at a time. Before we get into how this is done, observe that if we only work on one instruction at a time, a particular stage only performs work on one clock cycle of the machine cycle. For example, the Fetch Stage is active on the first clock cycle of the machine cycle. During the second, third, fourth, and fifth clock cycles, it is idle, which you might argue is wasteful. So, instead, after the first cycle, we might use it to fetch the next instruction, even before the current instruction is finished executing. Extending this simultaneous operation to all stages gives us the foundations of pipelining.

---

[5]When caching is being used the coupling of the Fetch Stage and the Memory Stage becomes even more complex. Caching of instructions by the Fetch stage can interfere with caching by the Memory Stage. That is to say, one stage would be replacing the cache entries loaded by the other stage, decreasing cache performance. It is for this reason that it is common for pipelined systems to have two cache units: a *data cache*, which is part of the Memory Stage, and an *instruction cache*, which is a part of the Fetch Stage.

TABLE 10.2 Eample 10.7 data hazard.

Cycle	Ftch	Dec	Ex	Mem	WtB
0	Add				
1	Mult	Add			
2		Mult	Add		

With the pipe architecture of Figure 10.4, we can work on five instructions at the same time. This is further illustrated with Table 10.1. Table 10.1 shows the five stages, and the instruction on which they are working, during eight clock cycles. During Cycle 0, the Fetch Stage is fetching the first in a sequence of instructions, called $I1$. The other stages are idle. During Cycle 1, $I1$ is being decoded, and a new instruction, $I2$, is being fetched. In Cycle 2, $I1$ has moved on to the Execute stage, $I2$ is being decoded, and the instruction $I3$ is being fetched. In Cycle 3, the instruction $I4$ is fetched, and the other instructions move on to the next downstream stage. In Cycle 4 another instruction, $I5$, enters the pipe, and the pipe has been completely loaded. Then, in Cycle 5 the pipe starts emptying, as instruction $I1$ finishes the Write-Back. In Cycles 5, 6, and 7, $I2$ finishes first, then $I3$ finishes, then $I4$, and then $I5$.

10.1.2.1 Problems with Pipelines

We have now covered the basics of pipelining. On the surface, it looks like a simple, elegant way of introducing parallelism, without some of the problems associated with a multi-processor system. And, in fact, it is. But, when we start looking a little closer, pipelining starts to show itself as a little messier than we originally might have thought.

There are two major problems that occur in pipe architectures.

- *Data hazards.* A result of an unfinished machine instruction is needed by a later machine instruction.

- *Branch hazard.* A result from a conditional branch instruction is needed to determine which instruction to fetch next.

Neither of these two hazards is catastrophic, however as stated, they make our original concept of a pipeline a little messier.

Let us do an example that demonstrates a data hazard. Consider the following assembly code snippet.

$$
\begin{array}{ll}
\texttt{add R0, R1, R2} \\
\texttt{mult R2, R0, R2}
\end{array}
\qquad (10.7)
$$

The first instruction adds the contents of two registers, R1 and R2, and stores the result in the register R0. The second instruction multiplies the result stored

TABLE 10.3   Example 10.7 with a 3-cycle stall.

Cycle	Ftch	Dec	Ex	Mem	WtB
0	Add				
1		Add			
2			Add		
3				Add	
4	Mult				Add
5		Mult			

in R0 by R2. Clearly, the result from the *add* instruction is used in the *mult* instruction. Table 10.2 shows how these two instructions traverse the pipeline, through the first few cycles. The point of Table 10.2 is that, in the last cycle shown, Cycle 2, the multiplication instruction is fetching one of its operands from R0, under the assumption that R0 contains the results of the addition instruction. However, the addition instruction, in Cycle 2, is just computing its result in the Execute Stage, and this result will not be written to R0, until two clock cycles later. If the operand for the multiplication instruction is read, the Decode stage will incorrectly read in an old value of R0. This is a data hazard; the result of one unfinished instruction, the *add* instruction, is needed by a later instruction, the *mult* instruction.

You might wonder how the data hazard problem can be solved. One easy solution is to introduce a *hardware stall* when a data hazard is detected by the processor. In a hardware stall, the processor waits several clock cycles before introducing the next instruction into the pipeline. Table 10.3 shows how a 3-cycle stall can be used on the machine of Figure 10.4, to solve the data hazard problem of Example 10.7. A 3-cycle "bubble" has been introduced between the addition and multiplication instructions. The effect is that when the multiplication instruction reaches the Decode Stage, the addition instruction has completed the Write-Back stage, and R0 contains the correctly updated value. Although the stall fixes the data hazard, it slows down the pipeline, allowing three of the stages to be idle. This diminishes the advantage of pipelining.[6]

Turning our attention to branch hazards, these are hazards that can be caused by any conditional branch instruction. Consider the following branch

---

[6]The hardware stall is not the only way to address the data hazard problem. In another solution, called a *short-circuit*, the Decode Stage can collect its operands, not only from memory or registers, but also directly from the output of the Execute Stage, allowing it to obtain the result even before it is written back. In another solution, called the *software stall*, the user is required to stall the pipe, by inserting several *nop* (no-operation) instructions, between the two arithmetic instructions. The *nop* instruction is an instruction that does nothing. Often, since machine language is ultimately produced by a high-level language compiler, it is the compiler that takes care of inserting these instructions, automatically.

TABLE 10.4   Example 10.8 branch hazard.

Cycle	Ftch	Dec	Ex	Mem	WtB
0	Beq				
1	I1	Beq			
2	I2	I1	Beq		
3	I3	I2	I1	Beq	
4	I4	I3	I2	I1	Beq

instruction.

$$\texttt{beq R0, R1, xyz} \tag{10.8}$$

This instruction branches to the address xyz, if R0 equals R1, and continues in sequence if not. Table 10.4 shows how this instruction traverses the pipe. The problem here is that by the time the *beq* instruction writes its result to the PC in the Write-Back Stage, and does the branch, the next four instructions in sequence, I1, I2, I3, and I4, have been started. If the branch is taken, these four instructions should never have been started; the instructions at address xyz should have been started.

Dealing with a branch hazard, again, can be done using a hardware stall. For our 5-stage architecture, if whenever we are executing a conditional branch, we stall for four clock cycles before inserting the next machine instruction into the pipe, after the stall, the PC will be updated correctly, and we will insert the correct instruction into the pipe. The problem with this stall strategy is that, for every conditional branch, we end up emptying the pipe, and greatly diminishing its effectiveness.[7]

10.1.2.2   Pipeline Performance

We have presented pipelining as a performance enhancer. So, the next question to consider is how much performance improvement can we expect? To start answering this question, we need to spend some time talking about how we measure the performance of a pipeline system.

---

[7] Other ways of handling the branch hazard are used. In a software stall, the compiler inserts *nop* instructions after a conditional branch. This is not much better than the hardware stall. In a second scheme, called *instruction annulling*, the processor goes ahead and starts more instructions through the pipe, not really knowing whether it should, and if the instructions are erroneously started, at the Write-Back stage, their results are discarded, rather than being written. This has the effect of canceling the instruction. In conjunction with annulling, *branch prediction*, predicting what a conditional branch will do based on recent history, is useful. This can diminish the chances that the processor will have to annul several instructions, increasing the effectiveness of the pipeline. A variant of a software stall is what is called *instruction reordering*, in which instead of inserting *nop* instructions after a conditional branch, other instructions from the program, that are independent of the branch, are used to fill the bubble. This has the advantage over a software stall of keeping all stages of the pipe occupied with useful work.

When we measure performance of a pipeline system, we are usually interested in comparing it to a system without pipelining. And, we are usually interested in how much faster we can execute an instruction, with pipelining. We have already explained that we are not directly increasing instruction speed, but rather increasing throughput. Increasing throughput, however, affects the average instruction speed on the machine.

To begin our discussion of performance, let us calculate the time it takes to execute a sequence of n instructions, on a k-stage pipeline machine. The expected time in clock cycles is the following.

$$T_k = (n - 1) + k \tag{10.9}$$

This formula comes from analyzing the nth instruction in the sequence. If we are starting a new instruction every clock cycle, the nth instruction waits $n-1$ clock cycles to enter the pipe, as each preceding instruction, in turn, enters the pipe. It then takes k cycles for the nth instruction to clear the pipe, resulting in the formula for T_k given in Equation 10.9. Notice that this formula for T_k assumes that the pipeline is operating at full capacity. While this assumption is probably not completely accurate, because of the hazards discussed in Section 10.1.2.1, it, at least, gives us a best-case estimate of performance.

Next, let us consider a machine that has the same stages, but no pipelining. For this machine, an instruction enters the pipe, exits the pipe k cycles later, and no other instruction is started while that instruction is in the pipe. The time to execute n instructions, for this machine, is the following.

$$T_1 = n \cdot k \tag{10.10}$$

We use the times T_k and T_1 to define performance.

$$P_k = \frac{T_1}{T_k} \tag{10.11}$$

This is the ratio of the expected times for the pipelined machine and the non-pipelined machine.

To demonstrate how we would use this measure of performance, let us look at an example. Suppose that we have a 4-stage pipeline machine, and we are executing 1,000 instructions. Then $T_k = (1,000 - 1) + 4 = 1,003$, and $T_1 = 1,000 \cdot 4 = 4,000$. The performance of the pipelined machine is then the following.

$$P_k = \frac{4,000}{1,003} \approx 3.988 \tag{10.12}$$

In other words, the pipelined machine is approximately four times as fast as the non-pipelined machine. You should see from this example that pipelining is an extremely powerful tool for increasing the performance of a processor, even if the pipeline is not always operating at full capacity.

10.2 INCREASING MEMORY SPACE

Remember that to execute a program it must be loaded into main memory. The question that then comes up is what happens if the program is so large that it cannot fit into main memory? This was often a significant problem, not too far in the distant past. Today, we hardly ever worry about this problem. Partially, this is because computer memories have become very large, but this does not solve the problem entirely, because as memory sizes increase, so do the sizes of programs that programmers produce. Possibly, the really significant advancement has been the development of computer systems with *virtual memory*. As an important advancement, virtual memory is the subject of this section.

10.2.1 The Memory Hierarchy

When we discussed cache in Section 10.1.1, we introduced the concept of having not just one, but several memory units. These several memory units duplicate each other. For instance, the cache contains entries that are also found in the main memory unit. Although we introduced cache to speed up memory fetches, we can easily extend the concept of duplicate memories to increase the size of the memory address space.

In Figure 10.5 we give a sample memory hierarchy. Starting at the top of Figure 10.5, we have the register file, inside the processor. The ultimate goal of the processor is to fetch data, or instructions, from memory, and store them in registers. Below the register file, is a cache unit. You may notice that the cache unit is labeled as an *L1* (level one) cache. The diagram also has an *L2* (level 2) cache. Data stored in registers are fetched from the L1 cache, if possible. However, if there is an L1 cache miss, then the processor would look for the data being fetched in the L2 cache. The L2 cache is a larger, and, most likely, a slower cache unit. If the fetch attempt results in an L2 cache hit, then the data would be fetched from the L2 cache, loaded into the L1 cache, and from the L1 cache, it would be loaded into registers.

When a fetch results in an L2 cache miss, with the architecture of Figure 10.5, the processor would then look for the data in main memory. If this resulted in a hit, the data would be loaded into the two caches and the register file.

So far, Figure 10.5 is simply describing the workings of a cache system. The only difference between this system and the direct mapped system that we

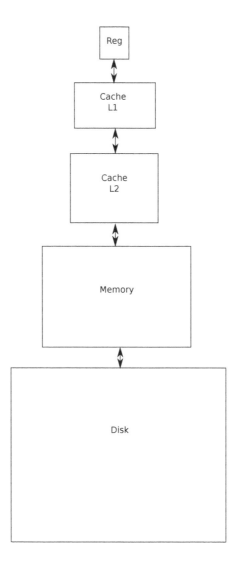

FIGURE 10.5 Hierarchy of memory devices.

described in Section 10.1.1.1 is that this system has a two-level cache system, which, hopefully, will speed up the processor just a little more than a one-level system. But, the bottom of Figure 10.5 introduces the way in which we extend memory.

Remember that to execute a program, it must first be loaded from disk into memory. The key observation that allows us to extend our memory space is that we only need to load the part of the program where we are currently executing, into memory. To illustrate this concept, suppose that we are executing a program of size 4 Mb, but our memory unit is of size 1 Mb. To enable the execution of the program, we split the program up into four equally sized pieces. We start by loading only the first 1-Mb piece of the program. We start executing this, and when execution moves out of this piece, into another piece, we load in the new piece. If any changes were made to the old piece, we write it out to disk, before it is replaced. This "piece swapping" process is represented in Figure 10.5 as the bottom part, where main memory is being fed by the disk.

10.2.2 Virtual Memory

Using disk as an extension of memory is an implementation of what is referred to as *virtual memory*, indicating that although memory appears to be larger, the main memory unit itself is not really all that large. There are two commonly used virtual memory schemes, that differ in how the program is split into pieces: *paging*, and *segmentation*. In paging, we split the program up into equally sized blocks, whereas in segmentation, we split the program up into blocks of varying sizes. Each of these two schemes has its own advantages and disadvantages, which, because of the condensed nature of this book, we do not have time to discuss fully. Instead, we will concentrate our discussion on paging, which is a fairly common virtual memory system.

10.2.2.1 Paging

As already explained, in paging we split a program that is too large to fit into memory, into fixed-sized blocks, called *pages*. The pages of the program are loaded into memory as needed, while the program executes. This paging activity is carried out by a combination of hardware, and operating system code. Probably it is best to illustrate this process with a simple example

Let's suppose that you have a memory unit, with a length of 4K words. You are trying to execute a program that occupies 16K words. With a 4K memory unit, since $4K = 2^2 \times 2^{10}$, the address would be 12 bits. This address will be referred to as the *physical address* of the machine. Now, let us imagine that the program is loaded into a "virtual memory unit," to be executed. This unit would have to have a length of at least $16K = 2^4 \times 2^{10}$. The address for the imaginary unit would then be 14 bits. We will refer to this address as the *virtual address* for the machine.

To implement the virtual memory, we will use paging. We decide that a page will be a block of size 1K words.[8] Notice that memory is large enough to contain four pages, simultaneously. The program is split into sixteen pages, and so only one quarter of them fit into memory, at a given time.

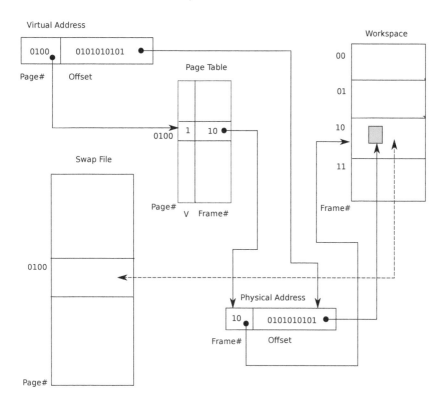

FIGURE 10.6 Address translation for paging.

In Figure 10.6 we show how pages are loaded into memory. On the right side of the diagram is a block marked as a workspace. This is the portion of the physical memory unit allocated to the program. It is split into four slots. Each of these slots is referred to as a *frame*, and the frames in Figure 10.6 are

_____

[8]The size of a page is not arbitrary. It involves a trade-off. With excessively small pages, during the execution of a program, many pages will be fetched from disk. Each disk access will slow down the execution of the program. So, it is beneficial to choose a large page size, to decrease the number of disk accesses. On the other hand, if the page size is too large, fetching a single page will take a long time, again causing the program to slow down. A compromise page size is needed, which keeps down the number of disk accesses, but also keeps down the read time for a single page.

numbered from 00 to 11 in binary. Each frame contains one of the pages from virtual memory, or it could be empty. We often refer to a workspace, set up in this fashion, as a *frame table*.

On the bottom left of Figure 10.6 is a block representing the *swap file*. The swap file is a file on disk, containing the running program. So, for our example, this would be a file with a length of 16K words. It is split into sixteen 1K-word pages, and each page has been given a page number, from 0000 to 1111, in binary (0 to 15 in decimal). The swap file represents the virtual memory for the program.

On the top-left of Figure 10.6 is a diagram of the virtual address for our machine. Remember that it contains fourteen bits. It has been split into two fields. The leftmost four bits are the page number, and the low-order ten bits are an offset. The page number of an address specifies what page the address is in, and the offset gives the number of words, from the beginning of the page to the word designated by the address. In the example shown in Figure 10.6, the virtual address is 01000101010101, which splits into a page number of 0100, and an offset of 0101010101.

When accessing a virtual address, like that shown in Figure 10.6, the corresponding page must be brought into RAM. Actually, it is possible that the corresponding page is already in memory. Either way, whether the page is in the frame table, a condition called a *page hit*, or the page is not in the frame table, a condition called a *page fault*, to understand the process of loading a page, there are several questions that need to be answered.

1. How do we tell if a page is already in memory, or not?

2. How do we translate the virtual address into the physical address?

3. What must be done to pull a page in from the swap file, and put it in the frame table?

We start by examining Questions 1 and 2. Referring back to Figure 10.6, we direct your attention to the middle of the diagram, and to a block labeled as the *page table*. The page table is a structure that is located in a part of memory that is owned by the operating system. It is used to translate virtual addresses into physical addresses. This table contains an entry for every page in the program, and is indexed by page number. So, you can see in Figure 10.6, the page number field, from the virtual address, being used to access the page table. The entry from the page table contains at least two fields: a V (valid) bit and a frame number. The V-bit has the same function as the V-bit used in cache units; it indicates the presence or absence of a page in the frame table. We can now answer Question 1. To determine if a page is loaded in the frame table, look the page number up in the page table, and examine the V-bit.

Now, to Question 2: Suppose that we consult the page table, and determine that we have a page hit. If the page is in the frame table, its page table entry will tell us in which frame it has been loaded. This is given by the frame number field. We now have enough information to construct the physical address that corresponds to the virtual address. This is shown on the bottom-right of Figure 10.6. To construct the physical address of the given virtual address, we simply combine the frame number, from the page table, with the offset, from the virtual address. For the example of Figure 10.6, with a frame number of 10, and an offset of 0101010101, we get the 12-bit physical address of 100101010101. Once we know the physical address of the desired word, we can access the word in the frame table. We use the frame number to index the frame table, and then use the offset to access the desired word, inside the frame.

We now know how to use the page table to process a page hit. Question 3 has to do with page faults. So, the question is what is done if the CPU is fetching using a virtual address, and it is determined that the corresponding page is absent from the frame table? The process for handling a page fault starts the same way as the process for handling a page hit. That is to say, we begin by looking up the page number in the page table. Now, however, we find that the V-bit is clear, indicating that the page is not in the frame table. What must happen now is that the page must be loaded into the frame table. If one of the frames is empty, this is not a problem. We would simply read the page from the swap file, load it into the empty frame, and update the page table. To update the page table, we would set the V-bit for the requested page, and write in the frame number for the frame in which we placed the page. As an example, suppose that Page 0100 is requested, and we discover that Frame 10 is empty. We would set the V-bit in the page table for Page 0100, and write a 10 into the frame number field of the same entry.

For a harder question, what would happen if a page fault occurs, but the whole frame table is full? The answer to this question is probably obvious; we would have to remove one of the pages from the frame table, replace it with the new page, and update the page table. So, for example, if we request Page 0100, this results in a page fault, the frame table is full, we somehow choose Frame 10 to be replaced, and Frame 10 is currently occupied with Page 1001. We would then write out Page 1001 from the frame table to the swap file, clear the V-bit for Page 1001, read Page 0100 from the swap file into Frame 10, set the V-bit for Page 0100, and set the frame number for Page 0100 to 10. The process of replacing one page, in memory, with another page, on disk, is called a *page swap*. For the swap, writing out the replaced page is only necessary when we have written data onto the outgoing page while it was in memory, and it ensures that the corresponding page in the swap file remains consistent with the page being removed from the frame table.

10.2.2.2 Page Replacement

When a page fault occurs, and the frame table is full, we previously stated that we somehow find a frame to swap out. This raises the question of how to choose a page to swap out. One, seemingly silly, solution is to "flip a coin." That is, we might just randomly choose a frame to replace. Although this seems to be an ill-suited strategy, in terms of performance, it is not all that unreasonable. But, there are other strategies that are more intelligent.

Page replacement strategies in common use, including the random choice strategy just discussed, are listed below.

- *Random (RAN) replacement.* Randomly choose a frame to replace.

- *Least recently used (LRU) replacement.* Replace the page that has not been accessed for the longest time.

- *First in, first out (FIFO) replacement.* Replace the frame that has been sitting in the frame table the longest.

LRU and FIFO replacement make an attempt to remove the least useful page from the frame table. LRU considers a page less useful if it has been sitting idle for a long period of time. FIFO considers a page less useful if it has been in the frame table for a long time, regardless of whether it has seen access activity or not. RAN makes no attempt at all to choose a least useful page.

The choice of replacement strategy can affect the performance of the processor. Obviously, processor performance is degraded by simply implementing virtual memory; on a system in which the whole program fits into the workspace, you will not need to do any paging, and this will result in faster execution times than for a system that implements paging, where you must access disk throughout the execution of the program.[9] But given that this slow-down will occur, as a designer, you would like to minimize the slow-down by choosing a good replacement strategy. The problem, though, is that what constitutes a good replacement strategy for one program, may turn out to be suboptimal for another program. And in fact, with both LRU and FIFO

---

[9]Speeding up paging, is a major concern. A feature that is often implemented is to use an associative cache unit, called a translation lookaside buffer (TLB), for access to the page table. The TLB stores frame numbers for recently used pages. The translation process can then use the frame number in the TLB, instead of accessing memory, and consulting the page table.

replacement, it is possible to construct programs that result in excessive numbers of disk swaps. On the other hand, RAN replacement is mostly immune to these worst-case programs, although its best-case performance is not as good as LRU and FIFO. This is why, although not all that intelligent a replacement strategy, RAN is still attractive as an option.

10.2.2.3 Disk Access

It is obvious that when paging, minimizing the number of swaps is desirable. Each swap requires a disk access, which blocks the CPU from proceeding, until the access is complete.

An advantage of paging, for implementing virtual memory, is that dividing a program into pages can be done automatically. For instance, if your program is 16K words long, and your page size is 1K words, you simply mechanically split the program in 16 consecutive pages. But, this automatic sectioning can cause larger amounts of disk accesses during execution, because the locations of the page boundaries are arbitrary, with respect to the pattern of instruction execution.

To explain the problem with patterns of execution, suppose that we have a loop in a program, that is split into multiple pages. We would like the loop pages to remain in the frame table, as long as control remains in the loop. This would minimize disk accesses, as control moves through the loop. But, we have little control over whether the loop pages do, in fact, remain in the frame table, or not. And so, we may be swapping loop pages in and out of memory, for as long as control remains in the loop.

An alternative way of implementing virtual memory is a scheme called *segmentation*. In segmentation, a program is split into *segments*. A segment is a block that is of variable size, unlike a page which has a fixed size. Segmenting a program is typically not automatic; segment boundaries can be arranged to accommodate the pattern of execution of the program. In our loop example, we could arrange our segmentation so that the whole loop is contained in the same segment, which would reduce the number of disk accesses during loop execution.

Although using segmentation can accommodate execution patterns better that paging, it has its own problems. Making effective use of the workspace, with variable sized blocks can be challenging. For example, when loading a segment of size 2.5K words into the workspace, you may be replacing a segment of size 3K words. This would give you the 2.5K words needed by the new segment, but you would also be left with 0.5K words of unused space. The effect of having unused space in the workspace is that you will be storing less of your program code in memory, increasing the chances of segment faults in the future.[10]

10.2.2.4 Memory Protection

One of the beauties of virtual memory systems is that they, automatically, supply memory protection. A common worry is that one running process might reach out to another process's workspace, modify it, or read its contents. This constitutes a significant security risk. In the extreme, it can lead to malware, with the ability to erase any security measures implemented by the operating system, allowing it to commandeer the machine.

On a system that has no memory protection, there is nothing stopping a program from using any legitimate address, whether the address is in the program's own workspace, or not. Of course, as discussed in Section 6.1.3.4, there is protection that keeps a user program from accessing kernel code, but it is difficult to extend this to protect other users.

In a system with virtual memory, notice that the virtual memory space only spans the swap file of a particular program. What this means is that, as with the system without virtual memory, the program is allowed to access any legitimate address, but all legitimate addresses are interpreted as virtual addresses, and therefore can only access the program's own workspace.

This system of protection is fairly robust. Its weakness is the address translation mechanism, which translates a virtual address into a physical address. If compromised, the user's program could, conceivably, access a physical address which is not in its workspace. In particular, if the user changes the frame number in a page table entry to a number outside of the range of the frame table, the program could access another workspace. As a consequence, it is vital to protect the page table from the user, and allow changes to it by the operating system only, through a system service interrupt, as discussed in Section 6.1.3.4. And so, we find that the page table is always located in a segment of memory, protected, and accessible only in kernel mode.

10.3 SUMMARY

This book has covered a great deal of material. By no stretch of the imagination, however, is this all there is to say about computer organization and computer architecture. In this, the last chapter, we have covered a few more advanced topics, in a cursory fashion. We discussed caching, which allows us

---

[10]Paging and segmentation each have advantages and disadvantages. It is possible to combine segmentation and paging in a compromise system, in which a program is split into segments, and each segment is further split into pages.

to keep useful instructions in the processor. Then, we took a look at pipelining, that allows the processor to work on several instructions at a time. A last topic we examined was virtual memory, that allows us to run programs that are too large to fit into memory.

There are several topics that we find interesting, but it is hard to justify including them in a text as focused as this one. In parting, we list a couple of them, and encourage you to research them on your own.

10.3.1   I/O Structure

We did not spend much time discussing how the processor communicates with I/O devices. In Chapter 5 we used a very simple model to explain the communication process at a very high level.

One advantage of our simple communication scheme was that to send or receive information from our I/O device involved only a single transaction. Our unit of transfer was the computer word, and the bus we used was wide enough to transfer the word in a single bus operation. Often the amount of data sent to a device is too large to be sent in a single transfer. In this case, the data is transferred in a sequence of bus transfers, one right after another. This type of communication is called *serial communication*, as opposed to *parallel communication*, in which all bits are sent over the bus simultaneously. Serial communication often requires a method of synchronizing the processor and the device. Often this synchronization is implemented by sending control signals between the processor and the device, using what is called a *handshaking protocol*.

A topic that bears further examination is the way in which interrupts are processed. An interrupt causes the processor to enter into an *interrupt cycle*, which is a cycle, like the machine cycle, that transfers control to the interrupt service routine.

In a system that has interrupt-based I/O, it is possible for several interrupts, from several devices, to arrive at the CPU, simultaneously. To resolve this conflict, usually devices are assigned *interrupt priority* levels. The device with the highest priority would then be handled by the ISR, and the other devices would need to wait. This priority scheme ensures that devices that need immediate attention are handled first.

Another topic concerning I/O devices is transfer of data between an I/O device and memory. Memory is usually the ultimate source of computer output, or destination of computer input. In a common system design, the data transfer would involve the CPU registers as an intermediate source or destination. This means that the processor actively pulls or pushes data between the memory unit and the I/O device. To relieve the CPU from this tedious job, a device called the DMA (direct memory access) unit might be connected to the bus. In this architecture, the I/O device communicates with the DMA unit to affect the transfer, rather than the CPU. The DMA then transfers data into the RAM, while the CPU is free to do other tasks.

10.3.2 Parallel Architectures

In this chapter we, briefly, touched on introducing parallelism into computation. We discussed pipelining, and also mentioned systems with either multiple full processors, or multiple partial processors.

We did not have room to discuss many interesting issues concerning parallel architectures. In particular, we really did not spend time discussing how multiple processors communicate, and synchronize. The choice of communication strategies can have a profound effect on how well a parallel system performs. Poorly configured communication can create a bottleneck, that ends up with a situation where processors spend a lot of time waiting for each other. Well-configured communication can diminish the amount of time spent in synchronization.

We also did not have room to discuss how memory is handled on multiprocessor systems. It is possible to design a system in which all processors share the same memory unit. The problem here is that this memory unit can become a bottleneck, and there are problems synchronizing the processors. It is also possible to give each processor its own memory unit. Here, the problem is that the processors cannot see each other's work, and to use results from another processor, those results must be sent over some communication system.

A last design is to have one distributed memory unit. In this design, there is a single logical memory unit. It is implemented with several physical units, that are maintained by different processors. This scheme is a compromise between a single memory unit and multiple memory units. Although its advantage is that all processors can see the results of other processors, the problem is to keep the pieces of the distributed memory consistent with each other.

10.4 EXERCISES

10.1 Using the same machine specifications as are used in Figure 10.1, do the following.

a. Indicate which of the following memory references are hits and which are misses: 011100, 100001, 011101, 000111, 010000, 000111, 010000.

b. Draw a figure similar to Figure 10.1, showing the contents of the cache unit after these references. (Leave the data section blank.)

10.2 You have a memory unit of length 4K words. Your cache unit has 32 entries, with each entry containing 16 words. Draw a figure, similar to Figure 10.2, that shows the address format for the machine.

10.3 Compare the optimal performance of a 3-stage pipeline with a 5-stage pipeline, when executing 2,000 instructions.

10.4 The code sequence in Example P10.1 results in a branch hazard on the pipeline system of Figure 10.4. Assume that branch hazards are handled by the compiler, which inserts *nop* instructions in the code to insure that an incorrect instruction is never started.

$$
\begin{array}{ll}
& \texttt{add R0, R0, R1} \\
& \texttt{mult R0, R0, R2} \\
& \texttt{sub R0, R1, R0} \\
& \texttt{beq R3, R2, ovr} \qquad \text{(P10.1)}\\
& \texttt{add R3, R3, \#1} \\
\texttt{ovr:} & \texttt{mov R3, \#0} \\
& \texttt{mov R4, R3}
\end{array}
$$

a. Rewrite the code sequence with the *nop* instructions inserted.

b. How many clock cycles will it take to execute your solution from 10.4a, if the branch is not taken.

c. How many cycles will it take to execute solution from 10.4a, without pipelining, assuming that the branch is not taken?

10.5 This is like Exercise 10.4. For this exercise, do not insert *nop* instructions; rather, rearrange the existing instructions to ensure that no incorrect instruction is started. That is to say, you need to rearrange the code so that the execution of the three instructions following the branch is not dependent on the execution of the branch.

10.6 You have a system setup as shown in Figure 10.5. You are to calculate the expected time to fetch an instruction. The hit rate for the L1 cache is 65%, and for L2 it is 70%. The hit rate for the page table is 45%. Access times are 1 clock cycle for the L1 cache, 2 cycles for the L2 cache, 3 cycles to access the page table, 3 cycles to access the frame table, and 6 cycles to access the swap file.

10.7 The code of Example P10.2 is being executed on a machine that has a frame table of size two. The swap file has 3 pages. Page 0 contains this code, Page 1 contains the procedure P, and Page 2 contains the procedure Q. Assume that originally the frame table contains none of these pages.

$$
\begin{array}{l}
\texttt{for i = 1 to 3 do} \\
\quad \texttt{P()} \\
\quad \texttt{Q()} \\
\quad \texttt{P()}
\end{array}
\qquad (\text{P10.2})
$$

a. Calculate the number of page faults that occur if the replacement strategy is LRU.

b. Calculate the number of page faults that occur if the replacement strategy is FIFO.

Suggested Readings

[1] A. K. Agrawala and T. G. Rauscher. *Foundations of Microprogramming: Architecture, Software, and Applications*. Academic Press, New York, NY, 1976.

[2] S. D. Brown. *Fundamentals of Digital Logic with VHDL Design, 5th ed.* McGraw-Hill, New York, NY, 2009.

[3] S. D. Burd. *Systems Architecture, 7th ed.* Cengage Learning, Boston, MA, 2016.

[4] J. D. Carpinelli. *Computer Systems: Organization & Architectur.* Pearson, Boston, MA, 2006.

[5] C. N. Fisher, R. K. Cytron, and R. J. LeBlanc, II. *Crafting a Compiler.* Addison Wesley, Boston, MA, 2010.

[6] M. J. Flynn. *Computer Architecture: Pipelined and Parallel Processor Design.* Jones and Bartlett, Boston, MA, 1995.

[7] J. L. Gersting. *Mathematical Structures for Computer Science.* W. H. Freeman, New York, NY, 2007.

[8] M. Gorman. *Understanding the Linux Virtual Memory Manager, 1st ed.* Prentice Hall, Upper Saddle River, NJ, 2004.

[9] J. Handy. *Cache Memory Book, 2nd ed.* Morgan Kaufmann, Walthum, MA, 1998.

[10] J. L. Hennessy and D. A. Patterson. *Computer Organization and Design: The Hardware / Software Interface: 4th ed.* Morgan Kaufmann, Walthum, MA, 2012.

[11] K. R. Irvine. *Assembly Language for Intel-Based Computers, 4th ed.* Prentice Hall, Upper Saddle River, NJ, 2002.

[12] M. Jimenez, R. Palomera, and I. Couvertier. *Introduction to Embedded Systems: Using Microcontrollers and the MSP430.* Springer, New York, NY, 2013.

[13] G. Kane and J. Heinrich. *MIPS RISC Architecture, 2nd ed.* Prentice Hall, Upper Saddle River, NJ, 1991.

[14] D. W. Lewis. *Fundamentals of Embedded Software: Where C and Assembly Meet*. Prentice Hall, Upper Saddle River, NJ, 2002.

[15] Y. Li. *Computer Principles and Design in Verilog HDL*. Wiley & Sons, Hoboken, NJ, 2015.

[16] K. C. Louden. *Compiler Construction: Principles and Practice*. PWS Publishing, Boston, MA, 1997.

[17] K. C. Louden and K. A. Lambert. *Programming Languages: Principles and Practice, 3rd ed.* Course Technology, Boston, MA, 2012.

[18] M. M. Mano. *Computer System Architecture*. Prentice-Hall, Upper Saddle River, NJ, 1993.

[19] M. M. Mano and M. D. Ciletti. *Digital Design: With an Introduction to the Verilog HDL, 5th ed.* Pearson, Boston, MA, 2012.

[20] T. Ndjountche. *Digital Electronics, Volume 3: Finite-State Machines, 1st ed.* Wiley and Sons, Hoboken, NJ, 2016.

[21] L. Null and J. Lobur. *Essentials of Computer Organization, and Architecture, 4th ed.* Jones, and Bartlett, Burlington, MA, 2015.

[22] B. Parhami. *Computer Arithmetic: Algorithms and Hardware Designs, 2nd ed.* Oxford University Press, Oxford, UK, 2009.

[23] R. Paul. *SPARC Architecture, Assembly Language Programming, and C, 2nd ed.* Pearson, Boston, MA, 1999.

[24] A. Silberschatz, P. B. Galvin, and G. Gagne. *Operating System Concepts, 8th ed.* Wiley & Sons, Hoboken, NJ, 2012.

[25] W. Stalling. *Computer Organization and Architecture, 9th ed.* Pearson, Boston, MA, 2013.

[26] W. Stallings. *Operating Systems: Internals and Design Principles, 7th ed.* Prentice Hall, Upper Saddle River, NJ, 2011.

[27] A. S. Tanenbaum and T. Austin. *Structured Computer Organization, 6th ed.* Pearson, Boston, MA, 2013.

[28] F. Vahid. *Digital Design with RTL Design, VHDL, and Verilog, 2nd ed.* Wiley & Sons, Hoboken, NJ, 2010.

[29] F. Vahid and T. Givargis. *Embedded System Design: A Unified Hardware/Software Introduction*. Wiley & Sons, Hoboken, NJ, 2002.

[30] J. S. Warford. *Computer Systems, 2nd ed.* Jones and Bartlett, Boston, MA, 2010.

Index